T0319683

Conducting and Financing Low-carbon Transitions in China

Conducting and Financing Low-carbon Transitions in China

A Governmentality Perspective

Le-Yin Zhang

Professor of Urban Economic Development, The Bartlett Development Planning Unit (DPU), University College London, UK

Cheltenham, UK • Northampton, MA, USA

© Le-Yin Zhang 2021

All rights reserved. No part of this publication may be reproduced, stored in a retrieval system or transmitted in any form or by any means, electronic, mechanical or photocopying, recording, or otherwise without the prior permission of the publisher.

Published by
Edward Elgar Publishing Limited
The Lypiatts
15 Lansdown Road
Cheltenham
Glos GL50 2JA
UK

Edward Elgar Publishing, Inc.
William Pratt House
9 Dewey Court
Northampton
Massachusetts 01060
USA

A catalogue record for this book
is available from the British Library

Library of Congress Control Number: 2021946088

This book is available electronically in the **Elgar**online
Political Science and Public Policy subject collection
http://dx.doi.org/10.4337/9781788977395

ISBN 978 1 78897 738 8 (cased)
ISBN 978 1 78897 739 5 (eBook)

Printed and bound by CPI Group (UK) Ltd, Croydon, CR0 4YY

Contents

Tables

Exchange rates

Average annual exchange rate (RMB per USD)

Year	Rate	Year	Rate	Year	Rate
2011	6.46	2015	6.23	2019	6.90
2012	6.31	2016	6.64	2020	6.91
2013	6.19	2017	6.75		
2014	6.14	2018	6.62		

Source: NBS

Acknowledgements

It is hardly possible to acknowledge all the debts that I have incurred during the preparation of this book as so many people have lent support. I wish to thank especially the following organizations and individuals: my host institutions and funders as mentioned in the final part of Chapter 1 and my local hosts including Liu Guangzhu, Li Zengwu, Tu Qiyu, Yan Yanming, and Wang Yao; the numerous people who kindly arranged and granted interviews, gave me their time, and shared ideas and knowledge; Vanesa Castan Broto for having commented on parts of the manuscript; my family for love and care; and finally Alex Pettifer and his team, especially Kate Norman, at Edward Elgar for their patient support.

Abbreviations

ADB	Asian Development Bank
ALG	Advanced Liberal Government
AR	Administrative Rationalism
CAC	Command-and-Control
CAGR	Compound Annual Growth Rate
CAS	China Academy of Sciences
CBA	China Banking Association
CBEEX	China Beijing Environmental Exchange
CBI	Climate Bonds Initiative
CBRC	China Banking Regulatory Commission
CCD	Community of Common Destiny
CCDC	China Central Depository & Clearing Co., Ltd
CCP	Chinese Communist Party (also known as Communist Party of China)
CCPCC	CCP Central Committee
CCPCCGAO	CCP Central Committee General Affairs Office
CDM	Clean Development Mechanism
CEA	Carbon Emissions Allowance
CEDNE	China's Energy Development in the New Era (White Paper)
CER	Certified Emissions Reduction
CES	Cadre Evaluation System
CET	Carbon Emissions Trading
CFEN	China Financial and Economic News
CMTRS	Carbon Mitigation Target Responsibility System
COD	Chemical Oxygen Demand
COP	Conference of Parties
CRS	Cadre Responsibility System

CSRC	China Securities Regulatory Commission
CWPERCGHGE	Comprehensive Work Plan for Emissions Reduction and Controlling GHG Emissions
DESA	Department for Economic and Social Affairs
DRC	Development Research Center
EAO	Energy Administration Office
EBF	Extra-Budgetary Funds
EC	Energy Conservation
ECA	Energy Conservation Audit
ECAE	Energy Consuming Appliances and Equipment
ECAES	Energy Conservation Assessment and Evaluation System
ECEP	Energy Conservation and Environmental Protection
ECER	Energy Conservation and Emissions Reduction (*jieneng jianpai* in Chinese)
ECERAES	Energy Conservation and Emissions Reduction Assessment and Evaluation System
ECERSF	Energy Conservation and Emissions Reduction Special Fund
ECERTRS	Energy Conservation and Emissions Reduction Target Responsibility System
ECL	Energy Conservation Law
ECO	Energy Conservation Office
ECQ	Energy Consumption Quotas
ECTRS	Energy Conservation Target Responsibility System
EEI	Energy Efficiency Improvement
EEL	Energy Efficiency Labelling
EES	Energy Efficiency Standard
EFC	Energy Foundation China
EM	Ecological Modernization
ENGO	Environmental Non-Governmental Organization
EP	Equator Principles
EPAs	Environmental Protection Agencies
EPC	Energy Performance Contracting

ESG	Environmental, Social and Governance
ETS	Emissions Trading System
EU	European Union
EUA	Energy Use Allowance
EUUs	Energy Using Units
EV	Electric Vehicle
FA	Fragmented Authoritarianism
FAI	Fixed Asset Investment
FCP	Four Cardinal Principles
FI	Financial Institution
FiT	Feed-in Tariffs
FTZ	Free Trade Zone
FYP	Five-Year Plan
GBEPC	Green Bond Endorsed Project Catalogue
GBIG	Green Bond Issuance Guidelines
GCBSA	Green Credit Business Self-Assessment
GCF	Gross Capital Formation
GCG	Green Credit Guidelines
GCPE	Green Credit Performance Evaluation
GCSS	Green Credit Statistical System
GCSSS	Green Credit Special Statistical System
GDI	Green Development Index
GDP	Gross Domestic Product
GEV	Green Energy Value
GFB	General Fund Budget
GFC	Green Finance Committee
GFSG	Green Finance Study Group
GFTF	Green Finance Task Force
GHGs	Greenhouse Gases
GIGC	Green Industry Guidance Catalogue
GIO	Gross Industrial Output
GLCDTTP	Green Low-Carbon Development Think-Tank Partnership

GLF	Gross Capital Formation
GOA	Government Offices Administration
GPB	General Public Budget
GTTs	Governing/Governmental Techniques and Technologies
HDRC	Hangzhou Development and Reform Commission
HMBS	Hangzhou Municipal Bureau of Statistics
HMCCCP	Hangzhou Municipal Committee of the CCP
HMPG	Hangzhou Municipal People's Government
HMPGGAO	Hangzhou Municipal People's Government General Affairs Office
IB	Industrial Bank (China)
IBGFG	Industrial Bank Green Finance Group
ICBC	Industrial and Commercial Bank of China
IEA	International Energy Agency
IFC	International Finance Corporation
IIGF	International Institute of Green Finance
IISD	International Institute for Sustainable Development
INDCs	Intended Nationally Determined Contributions
IPCC	Intergovernmental Panel on Climate Change
IRENA	International Renewable Energy Agency
JDDRC	Jingan District Development and Reform Committee
KPIGCI	Key Performance Indicators of Green Credit Implementation
LCCP	Low-Carbon City Pilot
LCDP	Low-Carbon Development Plan
LCE	Low Carbon Economy
LCECCC	Low Carbon Economy and Climate Change Center (Shanghai)
LCEI	Low Carbon Economy Index
LCOE	Levelized Cost of Energy
LCT	Low-Carbon Transition
LGB	Labelled Green Bond
LTLCDSLCTP	Long-Term Low-Carbon Development Strategy and Low-Carbon Transition Pathways

MA&E	Monitoring, Assessment and Evaluation
MEE	Ministry of Ecology and Environment
MEES	Mandatory Energy Efficiency Standard
MEP	Ministry of Environmental Protection
MEUEs	Major Energy Using Enterprises
MEUUs	Major Energy Using Units
MHURD	Ministry of Housing and Urban and Rural Development
MIIT	Ministry of Industry and Information Technology
MLCT	Mitigation and Low Carbon Transitions
MLF	Medium-Term Lending Facility
MLP	Multilevel Perspective
MNR	Ministry of Natural Resources
MoF	Ministry of Finance
MPA	Macro Prudential Assessment
NBS	National Bureau of Statistics
NCCP	National Climate Change Programme
NCSC	National Center for Climate Change Strategy and International Cooperation
NDCs	Nationally Determined Contributions
NDRC	National Development and Reform Commission
NEA	National Energy Administration
NEPVs	New Energy Passenger Vehicles
NEVs	New Energy Vehicles
NFFs	Non-Fossil Fuels
NFI	Non-Financial Institution
NGFS	Network of Central Banks and Supervisors for Greening the Financial System
NHREEC	Non-Hydro Renewable Energy Electricity Consumption
NPC	National People's Congress
NPCC	National Plan on Climate Change
PBC	People's Bank of China
PBCRB	People's Bank of China Research Bureau

PEC	Primary Energy Consumption
PED	Primary Energy Demand
PEVs	Plug-In Electric Vehicles
PPPs	Private–Public Partnerships
PRC	People's Republic of China
PRCEE	Policy Research Center on Environment and Ecology
PV	Photovoltaic
QDRC	Qingdao Development and Reform Commission
QMBS	Qingdao Municipal Bureau of Statistics
QMPG	Qingdao Municipal People's Government
RCCEF	Research Center for Climate and Energy Finance
RDA	Regionally Decentralized Authoritarianism
RE	Renewable Energy
REDSF	Renewable Energy Development Special Fund
REEC	Renewable Energy Electricity Consumption
REL	Renewable Energy Law
REQ	Renewable Energy Quota
RMI	Rocky Mountain Institute
SASS	Shanghai Academy of Social Sciences
SBC	Soft-Budget Constraint
SCGAO	State Council General Affairs Office
SDRC	Shanghai Development and Reform Commission
SECERC	Shanghai Energy Conservation and Emissions Reduction Center
SEDPs	Socio-Economic Development Plans
SEEE	Shanghai Environment and Energy Exchange
SEIs	Strategic Emerging Industries
SEPA	State Environmental Protection Administration
SEPCR	Strategy on Energy Production and Consumption Revolutions
SFSG	Sustainable Finance Study Group
SGF	Syntao Green Finance
SHETIC	Shanghai Economy and Information Technology Commission

SMBS	Shanghai Municipal Bureau of Statistics
SME	Socialist Market Economy
SMPG	Shanghai Municipal People's Government
SNGs	Sub-National Governments
SNM	Strategic Niche Management
SOD	Scientific Outlook on Development
SOEs	State-Owned Enterprises
SPDB	Shanghai Pudong Development Bank
SSE	Shanghai Stock Exchange
SZSE	Shenzhen Stock Exchange
Tce	Tonne of Coal Equivalent
TCFD	Taskforce on Climate-Related Financial Disclosure
TEC	Total Energy Consumption
TIS	Technological Innovation System
TM	Transition Management
TNC	Transnational Corporation
TRS	Target Responsibility System
UN	United Nations
UNEP	United Nations Environment Programme
UNFCCC	United Nations Framework Convention on Climate Change
WCED	World Commission on Environment and Development
WGUKCCEIDP	Working Group of UK–China Climate and Environmental Information Disclosure Pilot
WMO	World Meteorological Organization
WPCGHGE	Work Plan for Controlling GHG Emissions
WRI	World Resources Institute
WTO	World Trade Organization
YoYG	Year-on-Year Growth
ZCCCLCDC	Zhejiang Center for Climate Change and Low-Carbon Development Cooperation

1.   The China phenomenon

From collapsing snowscapes in Antarctica, to raging wildfires in Australia, and devastating floods in South Asia and Africa, climate change is causing havoc across the planet and will continue to do so with increasing frequency and intensity. Future generations will surely ask why we did not do more for mitigation and low-carbon transitions (MLCT). By mitigation it is meant any human intervention to reduce the sources or enhance the sinks of greenhouse gases (GHGs), whereas low-carbon transitions (LCT) denote system-wide processes in which the carbon intensity of human activities is reduced very significantly. Yet from our standpoint today, obstacles seem to abound in every direction. Ignorance, misunderstanding, and inertia are among our greatest woes.

Until recently, one of the most disturbing facts about climate actions is that the nationally determined contributions (NDCs), which are at the heart of the Paris Agreement, the international climate treaty adopted in 2015, would not achieve the stated objectives of the Agreement even if they were fully realized (Christensen and Olhoff 2019). The overall goal of the Agreement is 'holding the increase in the global average temperature to well below 2°C above pre-industrial levels and pursuing efforts to limit the temperature increase to 1.5°C above pre-industrial levels'. Scientists point out that the current level of NDC ambition needs to be roughly tripled for emissions reduction to be in line with the 2°C goal and increased fivefold for the 1.5°C goal (WMO et al. 2019). The latter would translate into a fivefold increase in the rate of decarbonization, every year from now, at a historically unprecedented rate of 11.7 per cent per annum (PwC 2020). Thankfully, by March 2021, over 120 governments around the world had announced their intention to reach carbon neutrality by the middle of the twenty-first century in line with the 1.5°C goal (IEA 2021a). This new level of ambition is certainly to be celebrated. However, it is only the first step in a long arduous journey.

There is no doubt that the goal of carbon neutrality will have profound impacts on world development in the coming decades. Its widespread adoption has added a new urgency to understanding how to tackle MLCT more effectively. An often-underappreciated point is that the climate challenge is a developmental challenge at the same time. This is explicitly recognized by the United Nations Framework Convention on Climate Change (UNFCCC), which states:

> The ultimate objective of the Convention is to achieve … stabilization of green-house gas concentrations in the atmosphere at a level that would prevent dangerous anthropogenic interference with the climate system. Such a level should be achieved within a time-frame sufficient to allow ecosystems to adapt naturally to climate change, to ensure that food production is not threatened and to enable economic development to proceed in a sustainable manner. (UN 1992, p. 9)

In other words, the challenge of climate actions is not just to reduce emissions at the required speed, but to do so without seriously affecting economic development at the same time. It is worth pointing out that the relationship between mitigation and economic development is not a matter of 'either/or'. Mitigation is intertwined with economic development because energy is indispensable to economic activities. To avert the worst climate change, we must de-couple economic activities from GHG emissions by either substituting fossil fuels with clean sources of energy, or improving energy use efficiency, or both (DTI 2003). However, without economic development, the chance of de-coupling would be severely restricted due to the lack of new resources and the limited scope for technological upgrading, both of which are essential for MLCT. Therefore, effective climate actions need to include rather than preclude economic development. This makes China's experience especially interesting. The latter shows that it is possible to achieve rapid economic development while curbing GHG emissions growth.

RATIONALE OF THIS BOOK

Given the growing acceptance of the carbon neutrality goal and its symbiotic relationship with economic development in the coming decades, a critical question facing the world is: 'How can we make nation states more effective in pursuing MLCT objectives while developing?' A possible way of address-ing this question is to learn from those nation states that have exhibited such effectiveness. We suggest that China is one of these nation states for several reasons. First, since the middle of the 2010s, China has increasingly been regarded as a leader on climate actions. It has attracted labels such as 'a world leader on climate change issues' (Henderson 2018), 'a new low-carbon cham-pion' (Engels 2018), and 'a global clean energy champion' (Andrews-Speed and Zhang 2019). China has scored 'good' from 2016 and 'high' (the top quarter) since 2018 on climate policy in the Climate Change Performance Index published by Germanwatch (http://www.ccpi.org). Second, China is a leader among G20 countries in decarbonization. It ranks second on PwC's Low-Carbon Economy Index (LCEI) league table, which compares performance among G20 countries in reducing carbon intensity, measured by changes in tCO_2 per \$m of GDP per annum. It decarbonized by 2.9 per cent per annum over 2000–19, compared with a rate of 1.5 per cent per annum for the world as a whole and 3.7 per cent for the top performer, the UK (PwC 2020). Third, although by far the largest energy related CO_2 emitter with a global share of 28 per cent, China's emissions have registered a marked slow-down since the early 2010s, as shown in Figure 1.1. From 2011 onwards, both annual fossil fuel-related emissions and per capita emissions in China nearly pla- teaued. As a matter of fact, on average, China's carbon intensity fell annually by 4.8 per cent during 2011–15 (Weng et al. 2018) and 4.6 per cent during 2016–19 (PwC 2017, 2018, 2019, 2020).

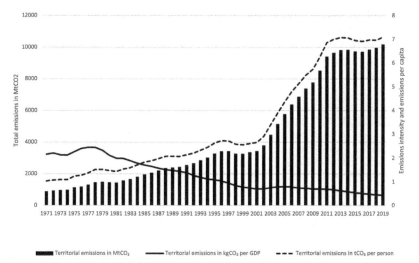

Source: Data from the Global Carbon Atlas (http://www.globalcarbonatlas.org) and Friedlingstein et al. (2020).

Figure 1.1 Levels of fossil fuel emissions in China, 1971–2019

Fourth, China has been relatively effective in fulfilling its climate objectives. Its carbon intensity fell by 48.1 per cent from 2005 to 2019 (State Council 2020), exceeding the Cancun Pledge a year ahead. According to the 2019 edition of the Emissions Gap Report (UNEP 2019), China was among the only three members of the G20 group, alongside the EU and Mexico, on track to meet their NDCs, which were considered reasonably ambitious. Moreover, China has steadily increased its climate ambition from its Cancun Pledge to its initial NDCs and the strengthened commitment announced in December 2020, although the latter is yet to be formally submitted (see Table 1.1). In addition, it has committed to achieving carbon neutrality before 2060. Fifth, China is by far the most active investor in MLCT technologies in the world. Between 2010 and 2019, China invested a cumulative total of USD 819 billion in renewable energy (RE) generating capacities, accounting for 30 per cent of the global total (State Council 2020). Sixth, while pursuing increasingly ambitious climate actions, China has managed to achieve rigorous economic growth. Between 2012 and 2019, while GDP grew by 7 per cent per annum, energy consumption grew by only 2.8 per cent per annum (implying an energy elasticity of 0.4) (State Council 2020). Seventh and finally, China has emerged as a major manufacturer and exporter of climate mitigation products and equipment (IEA 2018, p. 636). It accounts for a fifth of the world's exports in climate change mitigation technologies and is one of the top three countries in terms of low-carbon competitiveness in Asia (Srivastav et al. 2018). At the end of 2019, China produced 67 per cent of polysilicon, 79 per cent of photovoltaic (PV) panels, 71 per cent of PV modules, and 41 per cent of wind turbines in the world (State Council 2020).

Table 1.1   *China's Cancun Pledge and NDCs*

Cancun Pledge for 2020	2016 NDC for 2030	Revised NDC for 2030
1. To lower the CO_2 emissions per unit of GDP by 40–45% by 2020 compared to the 2005 level.	1. To peak CO_2 emissions around 2030 and making best efforts to peak early.	1. To peak CO_2 emissions before 2030.
2. To increase the share of non-fossil fuels in primary energy consumption to around 15% by 2020.	2. To lower CO_2 emissions per unit of GDP by 60% to 65% from the 2005 level.	2. To lower its carbon intensity by over 65% from the 2005 level.
3. To increase forest coverage by 40 million hectares and forest stock volume by 1.3 billion cubic metres by 2020 from the 2005 level.	3. To increase the share of non-fossil fuels in primary energy consumption to around 20%.	3. To increase the share of non-fossil fuels in primary energy consumption to around 25%.
	4. To increase the forest stock volume by around 4.5 billion cubic metres from the 2005 level.	4. To increase the forest stock volume by 6 billion cubic metres from the 2005 level.
		5. To increase the installed capacities of wind and solar power to over 1.2 billion kilowatts by 2030.

Source: Su (2010, 2015); Xi (2020).

Of course, China's MLCT performance in the past decade is not without problems. First, China is still among the bottom three of the G20 countries in terms of carbon intensity (PwC 2020). This demonstrates how much further China needs to go to decarbonize. Second, China's decarbonization so far has been mainly driven by improved energy efficiency, manifested in declining energy intensity, rather than large scale decarbonization of the energy system. Its carbon intensity index of energy, measured by CO_2 emissions per unit of total energy supply, has hardly changed since 2000 (IEA 2021b), although coal's share steadily declined from 68.5 per cent in 2012 to 57.7 per cent in 2019 (State Council 2020). This puts a question mark against the effectiveness of China's RE investment. Third, whether China can sustain its recent success is not entirely certain. In the most recent LECI league table by PwC (2020), China ranks the tenth among the G20 in reducing its carbon intensity in 2018–19, compared with first place in 2016–17, second place in 2015–16 and sixth place in 2017–18. China's carbon intensity fell by only 1 per cent in 2020 (NBS 2021), the lowest since 2011. Fourth and finally, China's NDCs are 'highly insufficient' in the sense that they fall outside of its 'fair share' range and are not at all consistent with holding warming to below 2°C according to the Climate Action Tracker (CAT 2021). Leading researchers in China estimate that to achieve the 2°C and 1.5°C goals would require China to reduce its energy-related emissions by 6.2 per cent and 8–10 per cent per annum between 2030 and 2050 respectively (He 2020). In other words, China would need to double its MLCT effort.

Thus, there is a distinctive duality to the China phenomenon. On the one hand, it has made significant progress in MLCT since the early 2010s. China's experience bears testimony to the possibility of substantially lowering carbon intensity without seriously disrupting economic development or requiring radical political change. On the other hand, China has evidently struggled with its decarbonization so far. This mixed picture raises questions about both how China has managed to achieve its de-coupling since the early 2010s and where China has encountered most difficulties in its decarbonization in the past decade. Answers to these questions will be useful to both China and other nation states as they embark upon a new and more ambitious round of decarbonization.

GAP BETWEEN THEORY AND REALITY

To some extent, existing studies using a modelling or extrapolation approach have provided some pointers as to what undergirds China's improved performance in the 2010s. These include faster than expected improvement in energy efficiency, the transition towards a more service-based economy, the development and deployment of clean energy technologies (IEA 2018), and slower economic growth and accelerated reduction in emission intensity (Garnaut 2014). Such analyses obviously represent important steps towards a fuller understanding of China's experience. However, we would like to argue that some of the factors, such as improved energy efficiency, economic structural change, and deployment of green energy technologies, are to a significant extent the outcome of state-led MLCT efforts so they need explanation themselves. This implies that we must look for more fundamental factors elsewhere.

However, a quick scan of the academic studies of China's climate actions shows that its recent improvement is largely unexpected and unexplained by social science research. Three examples suffice to illustrate this situation. Ploberger (2013) cautions us against thinking that, in the LCT, China would be able to repeat its earlier post-Mao success of adaptation towards a market economy through its reform programme. Deploying a structure-agency approach, he argues that, while the previous round follows both an economic and political logic, 'implementing a strategic shift towards a low-carbon economy will negatively affect the political and economic interests of various actors and consequently will generate resistance to such a change' (Ploberger 2013, p. 1042). Taking the lens of historical institutionalism, Andrews-Speed (2012) surmises that the nature of its institutions of governance will constrain China's adaptive capacity to move to a low-carbon economy (LCE). He suggests that accelerating the LCE transition will require 'significant or even radical institutional change across the polity, economy and society' in China (Andrews-Speed 2012, p. 219). Otherwise, he reckons that the CO_2 emissions might peak well after 2030 at 20,000 million tonnes or even higher. In a more recent publication, while conceding being too pessimistic in their early work, Andrews-Speed and Zhang (2019) argue that China's success in clean energy is despite the constraint of its governing institutions and at the expense of a large amount of political, human, and financial capital. Finally, Engels (2018) questions China's success so far and its chance of future success. Her argument is that whatever China has managed to achieve in mitigation so far is not a direct outcome of its responses to the climate challenge. Rather, 'it is more a side-effect of other domestic concerns than the outcome of a coherent strategic policy switch' (Engels 2018, p. 2).

The above examples reveal a state of common dismissal regarding China's MLCT performance. Given China's significant role in climate actions, such a state is problematic. As Green and Stern (2015) point out, China could potentially influence the course of global mitigation not only because of its sheer size and its influence in the developing world as a model of economic development, but also because China represents an important example in the task of tackling climate change (see also Green and Stern 2017). For the sake of protecting the future climate, we need to know more about China's experience, including how it has succeeded and where it has not. This book represents a humble contribution to a better understanding of the MLCT experience in China. It addresses three vital research questions: (1) How can we explain China's MLCT progress since the early 2010s? (2) What roles have state strategy, policy and programmes played? (3) What can other nation states learn from the Chinese experience about driving MLCT? It seeks to break new ground in three areas, including an integrated approach, a new theoretical lens, and the use of extensive primary data as well as secondary sources.

WHAT IS NEW ABOUT THIS BOOK?

In terms of scope, rather than focusing on specific aspects such as clean energy, this book aims to provide an integrated analysis of the Chinese MLCT experience. To learn lessons for nation states, we will endeavour to develop a holistic understanding of the regime of practices of the political authorities at different levels on the one hand, and across the different fields of economy, politics, and finance on the other hand. This book will also explore temporal changes in China's MLCT drive.

Furthermore, this book seeks to break new ground by introducing a novel theoretical lens. The literature review, presented in Chapter 2, shows that three mainstream theoretical perspectives on MLCT have so far strongly influenced the studies of China's climate actions. These are: the 'policy model' with emphasis on fixing market failures through public policy and institutions; the 'politics model' focusing on the politics of climate change; and the 'governance model' stressing the role of non-state actors in sustainability transitions. While their applications have generated some helpful insights and knowledge about China's MLCT, the extant studies seem to have worked better for highlighting what has not worked, rather than what has worked. To enable us to explore what has worked, we will adopt a different perspective or lens, that is, governmentality. This perspective capitalizes on Michel Foucault's pioneering work on discourse, the power/knowledge nexus and the art of government (e.g., Foucault 1991a, 1991c). The term 'government' here refers to the 'problematic of government in general' rather than 'the political form of government' (Foucault 1991a). It is about 'the conduct of conduct', or 'the way in which the conduct of individuals or groups might be directed' (Foucault 1982, p. 790). To Foucault, to govern means 'to structure the possible field of action of others' in a strategic way, rather than through violence and prohibition. It is achieved through governmentality, manifested in a series of strategies, technologies, and programmes of power (Gordon 1980). Foucault (1991b) points out that in studying governmentality, one is not 'assessing things in terms of an absolute against which they could be evaluated as constituting more or less perfect forms of rationality, but rather examining how forms of rationality inscribe themselves in practices or systems of practices, and what role they play within them' (p. 79). In the perspective of governmentality, it is hypothesized that China's improved MLCT performance since the early 2010s is explained by its effective carbon government, or the development and deployment of carbon governmentalities affecting the conduct of chosen target groups. Accordingly, our principal task in this book is to describe and explain these carbon governmentalities.

Four blind spots in the existing literature on MLCT in China become visible under the lens of carbon governmentality. First, existing studies have hardly explored the links between the broader state-led development discourses, notably the Scientific Outlook on Development (SOD) and the concept of 'Ecological Civilization', on the one hand, and the more specific discourses regarding MLCT on the other hand. Second, while some studies have alluded to the involvement of China's long-standing institutions such as the target-based performance system (e.g., Lo 2014) and five-year plans (FYP) (e.g., Yuan and Zuo 2011) in MLCT-related projects and programmes, few have attempted to examine systematically how these and other traditional governing techniques have been adapted, strengthened and combined with new ones to structure the field of action of the key subject groups around MLCT. Third, most studies adopt (often implicitly) a simplistic view of top-down Chinese politics and focus their attention on perceived central–local conflicts and implementation gaps, thus what does not work rather than what works. Such a view fails to take account of the considerable autonomy, capacities, and status of Chinese sub-national governments (SNGs) in managing local affairs, including MLCT. Given China's post-2006 system of 'management by region' for energy conservation and emissions reduction (ECER) (*jieneng jianpai* in Chinese) (Zhao and Li 2018), this constitutes a serious blind spot. Fourth and finally, while the world often marvels at the magnitude of China's investment in green industries and low-carbon infrastructure, there is hardly any effort to understand how such investment has been made possible at the level of financial governance. We intend to shed light on this aspect as an integral part of carbon government in China.

This book also strives to achieve a better balance between the use of primary and secondary data than what is usually available. As well as utilizing existing literature and available statistics, it will incorporate first-hand information gathered from the author's own fieldwork over 2013–18. Hopefully, this combination will help us to see more clearly what the various authorities in China set out to do, what governmental techniques and technologies (GTTs) they have developed and used, what they managed to achieve, and where the challenges and obstacles have been at both the central and local levels.

This book is structured as follows. Chapter 2 takes the first step in developing our understanding of the Chinese experience theoretically by providing an overview of the debates internationally on the role of different actors and means in pursuing sustainability and how this literature has influenced the studies of the Chinese experience. It also provides details of the Foucauldian governmentality perspective and finishes by operationalizing it for studying China's MLCT (see Table 2.1).

Chapter 3 then explores how China's MLCT rationalities are embedded in the discourses around SOD and Ecological Civilization and built upon a string of earlier discourses. It also explains how the governing subjects and objects are constituted and identifies eight key subject groups. Chapter 4 focuses on the GTTs that the Chinese state has adapted and utilized to affect the behaviours of the target groups. It identifies ten key elements in this set and highlights the importance of a tripartite governing mechanism comprising target-setting, monitoring, assessment and evaluation (MA&E), and sanctions for SNGs, major energy using units (MEUUs) and public officials. Chapter 5 examines the financing aspect of carbon government in China. It explores the relationship between the sources of MLCT investment and the foci of China's green finance governmentalities. In recognition of the urban nature of the LCT and the crucial role of urban authorities in China's MLCT, Chapter 6 investigates the experiences in three Chinese metropolitan centres (Shanghai, Qingdao and Hangzhou). This reveals a high level of diversity in terms of both the challenges encountered and the carbon governmentalities deployed. Chapter 7 concludes.

HOW THE BOOK CAME ABOUT

To set the book in its context, let me briefly explain the research processes that have led to the writing of this book. I started to develop an interest in climate actions from the early 2000s when I worked as a consultant for organizations including the UNFCCC and the UN Department of Economic and Social Affairs on the integration of climate change actions into national development strategies. Coupled with academic teaching and research activities based at University College London (UCL), these experiences have intensified my interest in the relationship between mitigation, LCT, and economic development. In 2013, funded by my home faculty, the Bartlett School of Built Environment, I had the first opportunity to conduct field investigations in Qingdao in terms of its MLCT experience. The choice of Qingdao was partly determined by its status as a low-carbon city pilot (LCCP), designated by the National Development and Reform Commission (NDRC) in 2012. It also allows me to draw on my prior knowledge of the city, gained from previous experience of supervising a doctoral dissertation about the science and technological parks in the city (Kim and Zhang 2008). Although my stay there, hosted by the Qingdao University of Sciences and Technology and the Government Offices Administration, lasted only a week, the visit enabled me to gain a first impression of MLCT processes in a Chinese city and to develop local contacts both inside and outside the local government. Then in 2015, with funding from the Urban Knowledge Network Asia (part of the European Union's 7th Framework), I used my secondment to the Shanghai Academy of Social Sciences (SASS), a think-tank, to investigate MLCT in Shanghai, another LCCP nationally designated from 2012. Having studied Shanghai's experience in economic development and its rise as a global city and the role of the state in these processes (Zhang 2003, 2014), I was keen to find out what role the state now plays in the city's MLCT process.

In 2017, with funding from the British Academy's International Mobility Scheme and in collaboration with the SASS, I spent two weeks in Shanghai and nearby Hangzhou, looking at how green finance was utilized by financial and other institutions to further MLCT. The city of Hangzhou is among the first batch of LCCP, having gained this status in 2010. It is where Alibaba has based its headquarters since its inception and where I previously lived and worked for two years in the early 1980s. That it is the capital city of Zhejiang province, a province where President Xi once served as the provincial Party Secretary (i.e., top leader) between 2002 and 2007, adds further interest. As a matter of fact, it was during his tenure in Zhejiang that Xi first propounded the idea that 'clear waters and green mountains are invaluable assets' (Xi 2007). Since 2012 and with his assumption of China's top leadership, this expression has become a motto throughout China whenever environmental issues are raised. Finally in 2018, as a visiting researcher at the International Institute of Green Finance (IIGF), Central University of Finance and Economics, I spent three weeks in Beijing, a city where I had lived for six years as an undergraduate and post-graduate student in the late 1970s and 1980s. While there, with the help of the IIGF and many kind contacts and friends, I explored the role of central policy makers and financial institutions (FIs) in the promotion of green finance in China. Altogether, this book is the culmination of 15 weeks' field investigation and many years' research.

I should also say some words about the data that this book has used. It draws on information collected from both desk research and personal fieldwork. Two observations are in order here. At a general level, information availability has vastly improved in China since the implementation of the Regulation on the Disclosure of Government Information from May 2008 (State Council 2007a). Many official documents are now made available online. This means that trawling government websites can unearth a great deal of information. However, 'sensitive' information such as that related to emissions remains hard to come by. Nevertheless, as the rest of this book will demonstrate, data triangulation has enabled me to piece together the jigsaw of China's MLCT experience and to appreciate some of the hitherto under-explored features and lessons.

2. Understanding China's low-carbon transitions in theory

Thomas Kuhn's theory of paradigms is a helpful reminder that what researchers try to do is to find a more effective way of understanding and explaining reality. It is in this spirit that I embarked upon the journey to understand why, despite the supposed unpromising conditions that it faces (see Chapter 1), China has managed to de-couple, at least partially, its economic growth from carbon emissions growth since the early 2010s. Could an alternative paradigm explain the China phenomenon better? This is the question that will be explored in this chapter. As it turns out, this exploration will take us into a wider field, namely the debate on the role of the state in the process of mitigation and low-carbon transitions (MLCT). This is inevitable and appropriate because, in the case of China, we are dealing with a socio-economic system that is dominated by state actors and their actions. Moreover, we are trying to extract lessons for other states.

Given China's importance as a major player in global MLCT, a voluminous literature has grown around its experience in this regard. Yet this literature seems to be limited on a crucial front: it does not adequately capture the central dynamic that has accompanied China's MLCT efforts, namely, the role of the Chinese state. More specifically, it fails to pay sufficient attention to the Chinese state's policy practices on the ground in promoting MLCT and their corresponding outcomes. To demonstrate this point and develop an appropriate analytical lens for the rest of this book, this chapter addresses two key research questions: (1) How have the mainstream theoretical perspectives on MLCT explained or failed to explain China's recent change? (2) Why and how could the governmentality perspective do a better job in explaining it?

This chapter is presented in two parts. It will first explain what the most influential theoretical models on MLCT entail and how they have been applied to China. The analysis will show that, despite the numerous insights generated, none of these perspectives has either anticipated or could explain the scope and speed of the transformation in China, largely because they fail to pay sufficient attention to the more thoughtful aspects of the practices of the public authorities regarding MLCT. Time lags between these studies and the recent nature of the MLCT progress are also a factor. The chapter will then explore why a governmentality perspective would help, before outlining how this perspective can be adapted to decipher the dynamics of the Chinese MLCT.

A TRINITY OF MODELS

The parable of the blind men and the elephant is well known. It is often used to illustrate squabbles between believers of different religions. There is certain similarity between the parable and what happens in social science research, except that, in the latter, the participants are not blind. Rather they are partial in their vision to the extent that they often deploy different perspectives or lenses. In this light, it is possible to identify three principal theoretical perspectives that have been applied to the studies of China's MLCT experience. The first focuses on addressing the market failures involved in climate change through public policy and institutions. The second one is preoccupied with the politics of climate change and LCT. And the third is mainly concerned with governance issues in sustainability transitions. I refer to them as the policy model, politics model and governance model respectively in this book. Each of these models has been applied to studies of China's MLCT experience in one way or another. We will review some of the more important applications below.

The Policy Model

Economists have had, no doubt, a long influence on transitions thought. Their insights coalesce around what we can call 'the policy model'. This model sees climate change as an extreme case of market failure, manifested through negative externalities. Key options of addressing climate change in this perspective then include fundamentally the introduction of a carbon price internalizing the negative externality, alongside the exploitation of energy efficiency improvement potential and the development and deployment of new technologies that can effectively reduce carbon emissions through public policy and institutional changes (Hood 2011). The emphasis on introducing a carbon price is meant to send a signal to firms and households and to stimulate research and development in low-carbon technologies. Related to that, climate action is conceptualized as a global public good. It means that there are no private incentives either for persons, groups, regions or countries to reduce emissions, hence the need for public intervention either at national or international levels (Stern 2007). Nordhaus (2006a) identifies three approaches to addressing climate change as a global public good: (1) command-and-control (CAC) regulations; (2) quantity-oriented market approaches; (3) tax or price-based regimes. He suggests that only the second and third would be efficient.

This model faces challenges on two principal fronts. On the one hand, creating an effective carbon pricing system has proved to be difficult in governance terms. In fact, carbon prices have declined, rather than risen, since the early 2010s. In 2020, the estimated global average carbon price was only USD 2/tCO$_2$ (World Bank 2020, p. 8). This is well below even what Nordhaus' (2006b) conservative estimates would require. Using the Dynamic Integrated Climate-Economy (DICE) model and a declining social discount rate of 3 per cent per year initially, Nordhaus derives an optimal carbon price of USD 17.12 per ton of CO$_2$ for 2005, rising over time to USD 84 in 2050 and USD 270 in 2100, or an optimal rate of emissions reduction of 6 per cent in 2005, 14 per cent in 2050, and 25 per cent in 2100, compared with the level of 1990. He shows that these figures would be consistent with a projected global temperature increase from 2000 to 2100 of around 1.8°C. On the other hand, a key solution, the development and deployment of cleaner technologies, suffers from market failures as well. The difficulty of appropriating the value of knowledge leads to socially sub-optimal private sector investments in LCT-related R&D, thus requiring policy to supplement with a 'technology-push' (Popp 2010). Moreover, uncertainty in the direction of technological shift and requirement for long-term investment further compound the difficulty (Mazzucato 2018). While the need to address financial market failures has motivated an international campaign to restructure the financial system (UNEP Inqury 2015), two different approaches to innovation policy have emerged, involving the correction of market failures in the technology market and the innovation system approach (McDowall et al. 2013). The former focuses again on promoting the formation of a carbon price, public subsidies for and investment in new clean technologies, and regulatory pressure for decarbonization. In contrast, the latter builds upon a wider understanding of technology and innovations. It sees technology as a socio-technical regime, whose transition is affected by both technological and social factors. For example, the technological innovation system (TIS) concept proposed by Jacobsson and Johnson (2000) suggests that, to be successful, a new radical innovation must be fostered within an innovation system that successfully performs a number of critical functions, or activities. The combined effect of these challenges is that the policy model based on carbon pricing is unlikely to be an effective means in the short and medium terms.

Existing China-specific studies in this area have focused on government policies supporting the acquisition, development and deployment of renewables technologies and the growth of related industries and their effects. There has also been significant attention to the carbon emissions trading (CET) pilots that China launched in 2013/2014. These studies show that, although there has been strong government policy intervention and widespread experimentation in these areas, such efforts are by no means consistent over time and unambiguously effective.

Adapting the innovation system approach, McDowall et al. (2013) explore the interaction over time between policy and innovation system dynamics and its effect on the development of wind energy in three jurisdictions: the EU, the USA and China. They find that, in China, the success of wind energy is better explained as an outcome of Chinese central government policy than as a result of international technology transfer measures such as the Clean Development Mechanism (CDM) or bilateral donor programmes. They also discover that the establishment of the legitimacy of wind as an important technology by China's central government is 'the key feature' in this success, which has been facilitated by international policy mechanisms such as CDM, international experience, and China's broader energy policy context. Also adopting the TIS framework, Huang et al. (2016) attempt to understand the rapid rise of the Chinese photovoltaic (PV) industry and its profound impact on the global PV industry and to explain the ingredients of this success. The authors use process tracing techniques to map out all the key events around the development of PV technologies and the PV manufacturing industry from 1985 to 2012. They conclude that the rise of the Chinese PV TIS can be explained by the interaction of three contextual factors, namely, the change in Chinese institutions, technology transfer and the large European market, and specific PV TIS dynamics. Finally, Mazzucato and Semieniuk (2018) look into the role of innovation policy and market creation in the growth of electric vehicles (EVs) in China. Acknowledging that China's share in total global EV stock has risen from 7 percent in 2009 to 39 percent in 2017, they attribute the success to both supply- and demand-side policy support, noting that subsidies of up to USD 8,736 per EV were renewed in 2016.

Other researchers have been more critical about the success of China's clean energy equipment manufacturing industry and policy support. Chen (2016) brands China's industrial policy on the PV industry as a failure because 'it only produced a large industry, not a competitive one' (p. 756). Moreover, Chen argues that this is rooted in the Chinese institutions where the central government sets the industrial policy and then leaves the local governments to implement the policy in a competitive fashion against each other. Also adopting the TIS framework, Zou et al. (2017) find that poor connectivity in networks, unaligned competitive entities and a lack of market supervision obstruct the development of China's PV industry. Wan et al. (2015) find that, despite large subsidies (up to USD 17,600 per vehicle in some regions such as Shanghai) and initial ambitions, the sale of EVs were meagre. In 2013, the sales of 17,600 plug-in EVs (PEVs) accounted for less than 0.1 percent of total civilian vehicle sales, compared with the central government's PEV sales target to reach 10 percent of automotive sales by 2012 nationwide (Wan et al. 2015).

In terms of creating a carbon price, this is still at an experimental stage in China despite its launch of seven regional CET pilots in 2013/2014 and its preparation for a nationwide CET platform since 2017. A study finds that these pilots had a total allowance of 1.2 billion tonnes of carbon dioxide per year, about 11.4 per cent of national emissions in 2014; from November 2013 to October 2016, the seven pilots traded 94 million tonnes of emission allowance at an average price of USD 3.72/ton (Zhang et al. 2017). The study points to various problems including a lack of legally binding commitment, largely free allocation of allowances, the lack of an over-arching design, market fragmentation and a lack of data transparency. Its key finding is that these pilots' impacts on emissions reduction, mitigation, cost-saving and air pollution are likely to be 'very limited'. Thus, while the policy model rightly places public policy at the heart of MLCT in China, its demonstratable effects on mitigation through the pricing mechanism are still limited. Moreover, these studies often point to non-economic contexts such as political prioritization and regional competition as more significant factors.

The Politics Model

Unlike 'the policy model', which is well established, what could be described as 'the politics model' is still in the making. This is because for a long time, politics was left out of much of the transition studies literature (Meadowcroft 2009), although there has been a growing consensus that politics is an important dimension of the process of LCT since the end of the 2000s (Meadowcroft 2011; Grin et al. 2010). Recently, serious efforts have been made to introduce the concept of power into transition studies (e.g., Avelino and Rotmans 2009). Attention is however focused on what is described as 'horizontal' rather than 'vertical' power (Avelino and Rotmans 2011, p. 800). In other words, the focus is on the conflicting roles of different actors, rather than the integrative role of nation states. Although Avelino and his co-workers follow Parsons (1963) by defining power 'as the ability of actors to mobilize resources to achieve a certain goal', they have excluded some key elements from Parsons' framework of political power. Thus while Parsons (1963, p. 237) defines power as the 'generalized capacity to secure the performance of binding obligations by units in a system of collective organization when the obligations are legitimized with reference to their bearing on collective goals and where in case of recalcitrance there is a presumption of enforcement by negative situational sanctions', the framework suggested by Avelino and his co-workers leaves out reference to collective goals and the connection between political power, binding obligations, collective goals, legitimization and sanctions. This means that the elaborate power framework proposed (Avelino and Rotmans 2009, 2011; Avelino 2017) is concerned with neither the state's mobilization of resources in the public interest nor the sanctions that might be imposed to solicit compliance. What is also left out is the notion of collective leadership, conceived by Parsons (1963, p. 255) as those 'who can mobilize the binding commitments of their constituents in such a way that the totality of commitments made by the collectivity as a whole can be enhanced'. In other words, as it currently stands, the politics model says literally nothing about how nation states might develop ambitious commitments to climate actions and how they can be effective in achieving their goals in the face of resistance.

The absence of a state-centred politics model for climate actions is unfortunate but is a logical consequence of the intellectual undercurrents since the 1970s. The latter include the rise of such ideas as 'governance without government' and the paradigmatic shift from statism to federalism and interactive governance (Peters and Pierre 1998; Elazar 1995; Torfing et al. 2012). However, it has serious consequences for climate change. Against the prevailing emphasis in politics on the enabling state, Giddens (2009) called for an 'ensuring state' in his influential book *The Politics of Climate Change*. He stresses that the state will be an all-important actor, 'since so many powers remain in its hands' (p. 4). He considers that the concept of the enabling state is not strong enough, because the state must ensure the desired outcome, that is, limiting global warming. Without the armoury of the state, such as planning, however, an ensuring state is hardly possible. Meanwhile, since the Copenhagen Summit in 2009, the focus of mitigation action globally has shifted from international treaties such as the Kyoto Protocol (UNFCCC 1997) to nationally determined, voluntary mitigation targets under the Cancun Agreement and the Paris Agreement (UNFCCC 2015), which effectively put nation states in the driving seat of international mitigation efforts. A key difficulty that this combination presents for those who wish to understand what the state could do for MLCT is that, with few exceptions (e.g., Barry and Eckersley 2005), there is an overall lack of theoretical attention to the positive role of nation states in tackling global ecological crises, including climate change. Indeed, the literature is dominated by a negative view about the role of state-centred politics. In his attempt to remedy the lack of politics in the influential multilevel perspective (MLP), Geels (2014), a key figure in developing the perspective, emphasizes resistance by incumbent regime actors to fundamental change. His essentially negative approach conceptualizes relations between policy makers and incumbent firms as a core regime-level alliance, which would resist fundamental changes required by LCT (Geels 2014, p. 26). In conclusion, this model does not lend itself to a positive analysis of state actors and actions. It fails to address two crucial questions: (1) Why would the state wish to ensure LCT? (2) How can the state help ensure LCT?

Extant literature on MLCT in China has examined various political factors and actors that have contributed to China's changing policy on climate change. These studies show that state policy has evolved along a path of increasing political commitment towards climate-related issues under the influence of both domestic and international factors. Stensdal (2012) identifies three key changes from the late 1980s. First, from 1988 to 1997, global warming was considered as an issue only for the developed countries and of limited interest to China. Second, from 1998 to 2006, global warming gradually became accepted as a national concern, albeit only as an environmental issue. Finally, from 2007, climate change has been elevated to the status of a national priority. Using the Advocacy Coalition Framework, Stensdal attributes these changes to the formation of an advocacy coalition, whose membership ranges from climate change scientists, media, officials of relevant government departments, some committed sub-national governments (SNGs) to businesses, the policy-oriented learning among the members and the sophistication of socio-economic conditions in China. By sophistication, Stensdal refers to the maturation of the relevant actors and growing Chinese expertise on matters relevant to climate change policies (Stensdal 2014). On the other hand, Huang et al. (2016) show that the Chinese central government responded to international signals such as the establishment of the UNFCCC by prioritizing the development of clean energy technologies and related industries. Moreover, they argue, this response in turn encouraged local authorities to provide support in terms of investment, low-cost land use and the like to these industries. They also point to the effects of the expansion of private ownership and the industry-specific lobby, something new to China. In the face of collapsing demand in Europe in the wake of the 2008 financial crisis, for example, the Chinese PV Industry Alliance joined forces with local authorities in successfully lobbying for a feed-in-tariff (FiT) scheme for PV panels. This led to the development of a domestic market for PV products. Through the lens of the theory of international relations, Zhang (2019) explores how China has emerged as a global leader in green finance partly through collaborating with and learning from international actors.

Many research efforts have been directed at assessing the central–local relationship over MLCT in China. For example, drawing on fieldwork results from Shanxi – a major coal-producing province in hinterland Northern China – in 2010, Kostka and Hobbs (2012) attest to a high level of local responsiveness to national guidelines regarding energy efficiency targets in the 11th FYP (2006–10). They highlight three specific implementation methods that local officials used to comply with central mandates: 'bundling' of the energy efficiency policy with policies of more pressing local importance; 'bundling' of energy efficiency policy with the interest of politically influential groups; and reframing of the policy as part of a new economic development model. This shows that some local officials did respond to the new agenda in a creative way. Moreover, based on interviews with executives at 43 Chinese wind and solar firms, Nahm (2017) argues that policy supports offered at different administrative levels have been complementary in supporting industrial upgrading and innovation in these industries.

The effects of Chinese authoritarianism on climate actions have been examined too. Branding the Chinese state as one of environmental authoritarianism, defined broadly as 'a policy model that concentrates authority in a few executive agencies manned by capable and incorrupt elites seeking to improve environmental outcomes' (p. 288), Gilley concludes that the Chinese system is good for policy output, for example meeting the target of carbon intensity reduction, but bad for implementation and outcome. His argument is that democratic environmentalism, defined as 'a public policy model that spreads authority across several levels and agencies of government, including representative legislatures, and that encourages direct public participation' (pp. 288–9), would have resulted in the identification of a more ambitious mitigation goal and more effective implementation. However, the Chinese government did propose a more ambitious mitigation policy goal in 2014 through its intended nationally determined contributions (INDCs) for the Paris Agreement, as discussed in Chapter 1.

On the other hand, influenced by the 'fragmented authoritarianism' (FA) model of Chinese politics, there is significant attention to the issue of the implementation gap (e.g., Chien 2013; Xu 2016; Westman and Castan Broto 2018; Shen and Ahlers 2018). Having grown out of research work by Lieberthal and Lampton (1992) in the 1980s, this framework asserts that 'policy made at the centre becomes increasingly malleable to the parochial organizational and political goals of various vertical agencies and spatial regions charged with enforcing that policy. Outcomes are shaped by the incorporation of interests of the implementation agencies into the policy itself' (Mertha 2009, p. 996). One of the key mechanisms underpinning this fragmentation is supposedly the *tiaokuai* system, which refers to contending power structures between those based on vertical sectoral lines by ministries and those on horizontal regional lines by provinces. The FA model has shown enduring influence in the field of environmental or LCT studies and tends to provide a ready explanation for policy failures. For example, Lo (2014) shows that only one of three low-carbon policies (the Ten-Thousand Enterprises Programme (2011), the Building Retrofit Programme (2008) and the Thousand Cars Programme (2009)) implemented in Changchun (a low-carbon city pilot (LCCP) in Northeast China) was successful in meeting centrally set targets. He then attributes this implementation gap to the lack of political interest at local level and some specific problems that the other two programmes suffered from. Lo (2014) concludes: 'In our view, given the power of local government leaders in controlling local functional departments vis-à-vis central ministries under the tiaokuai administrative system, their lack of political interest in low-carbon city initiatives goes a long way in explaining the implementation gap' (Lo 2014, p. 242).

However, it is possible that the FA model is ill-suited for explaining climate actions, which are deemed as political priorities. Edin (2003) argues that, thanks to the introduction of the Cadre Responsibility System (CRS), the state capacity – the capacity to monitor and control lower-level agents –increased in the 1990s in comparison with the 1980s despite decentralization. In her view, the CRS enables the centre to achieve selected policy goals that are deemed political priorities. On the other hand, Mertha (2009) shows that despite continuing authoritarianism, there is a trend of pluralization of policy making processes in China thanks to the emergence of what he called 'policy entrepreneurs' including: officials within Chinese government agencies opposed to a given policy, often because of official organizational mandates such as environmental protection; journalists and editors; and individuals within Chinese non-governmental organizations (NGOs). Elsewhere, Zheng et al. (2013) examine the political economy of the local cadres' incentive to tackle pollution issues. They find that changes in both the central directive and the local public's environmental awareness were driving mayors' environmental actions. Using a principal-agent framework and data from 86 prefecture-level cities, they assess statistically the effects of changing incentives, as reflected in the criteria of the target performance system and the strength of local public concern, on mayors' action in tackling pollution issues including the reduction of energy intensity. They conclude that both the central government and the public influenced the urban leaders to mitigate environmental externalities.

Overall, extant studies following the politics model help illuminate the growing level of political commitment, policy learning and actions towards MLCT amidst the pluralization of Chinese politics, the rising importance of the non-economic agendas and the importance of central–local relationship. However, under the influence of the FA model, they often fail to recognize the extent to which Chinese politics has pivoted towards low-carbon and green development since the mid-2000s, a trend that has accelerated since the early 2010s. On the other hand, extant studies have rarely attempted to examine the links between changing policy priorities, central directives, locally deployed mitigation strategies and mechanisms, and mitigation outcome. Time lag is also an issue: most of the studies published and reviewed above reflected more what happened during the 11th FYP, rather than the 12th and 13th FYP periods. As a result, these studies do not really engage with the question of why the state-led system in China has been able to put up a stronger MLCT performance since the early 2010s in terms of commitments and goal attainment, let alone explaining it.

The Governance Model

Governance is a somewhat loose term. It can either refer to all processes of governing (Bevir 2012) or any mode of coordination of interdependent activities (Jessop 1998). Therefore, governance includes both vertical mechanisms of coordination such as the bureaucracy and horizontal coordinative mechanisms such as market and social networks (Jessop 1998). The research on the efficacy, or the lack of it, of the CAC governance system fits in here. Cox (2016) has provided a synthesis of the so-called 'pathology of command and control'. He uses the term 'command and control' (CAC) to indicate 'a problematically large degree of authoritative centralization and control in a governance system, rather than a particular type of policy instrument (e.g., regulations instead of incentive-based instruments)' (p. 2). Cox (2016) identifies 21 elements in the theory of the pathology of CAC and groups them into four different sets of arguments or paths that allegedly make CAC pathological. These are: the enabling path, the panacea path, the suppression path, and the lock-in path. The overall claim is that the CAC type of governance system would facilitate the development and implementation of technical fixes and favour panacea or technical fixes that fit poorly with local context, thus leading to commons' degradation; meanwhile such fixes would suppress local variation and function and cause lock-in.

The term governance also has a narrower sense, as what Jessop (1998) calls 'heterarchy'. Governance in this narrow sense is characterised by horizontal relationships including interpersonal networks, inter-organizational coordination, and inter-system steering (Jessop, 1998). In fact, the usage of the term governance in the LCT literature tends to take this narrow sense. The preoccupation with horizontal relationships stems arguably from a critical reflection on policy practices in managing the environment and promoting sustainable development in Northern and Western Europe in the 1970s and 1980s (Hoogma et al. 2002; WCED 1987). It is this narrowly defined governance that defines what we call the 'governance model'. This model resonates with the 'governance without government' literature, with emphasis on networks, partnerships and markets (Peters and Pierre 1998), and the literature on interactive governance, defined as 'the complex process through which a plurality of actors with diverging interests interact to formulate, promote and achieve common objectives by means of mobilising, exchanging and deploying a range of ideas, rules and resources' (Torfing et al. 2012, p. 3). Furthermore, it rightly highlights the role of governance experiments and learning (Bulkeley 2010; Castán Broto 2017). However, in this narrow sense, governance 'is characterized by diversity, uncertainty, heterogeneity of society, and the *decreased possibilities for inducing long-term change by government*' (Loorbach 2010,

p. 166, added emphasis). In other words, the governance model emphasizes the possibility of effecting changes without the state's leadership.

The governance model has produced a large literature centred on three sub-frameworks: the MLP, Strategic Niche Management (SNM) and Transition Management (TM). A recent review of the relevant literature identifies a total of 115 related articles published in the preceding decade (Wieczorek 2018). Of the three sub-frameworks, the MLP is by far the most popular. It distinguishes three levels of analytical concepts for explaining technological transitions: niches, socio-technical regimes, and landscape (Geels and Schot 2007). Niches are protective spaces that facilitate experimentation with novelties, whereas the landscape refers to 'a broad exogenous environment that, as such, is beyond the direct influence of actors' (Grin et al. 2010, p. 23). Finally, socio-technical regimes are defined as 'relatively stable configurations of institutions, techniques and artefacts, as well as rules, practices and networks that determine the "normal" development and use of technologies' (Smith et al. 2005, p. 1493). In this perspective, a transition (such as LCT) is essentially the transition of the relevant socio-technical regime, which can only be realized through the alignment of the factors at the three different levels (Geels and Schot 2007). In contrast, SNM is focused on 'the managing of the process of (technological and/or market) niche creation, development and breakdown to enable regime-shifts' (Hoogma et al. 2002, p. 30), whereas TM focuses on the actions of frontrunners in a society that promotes sustainability (Loorbach et al. 2015).

With deep historic and theoretical roots and a diversity of sub-frameworks to choose from, the governance model is undoubtedly a very influential analytical framework. However, interactive governance practices often fail to deliver on climate action ambitions because of a lack of capacities of social actors and insufficient attention to political economy and to issues of power and authority (Bulkeley 2010; Newell et al. 2015). Moreover, proponents of this perspective often have a penchant for niches and have a dislike of technical and cultural solutions, as they view technology as part of the problem and culture as lacking in coordination and insufficient on its own (Hoogma et al. 2002). On the other hand, coupled with the emphasis on non-state actors and forces, researchers following this tradition are dismissive of the pathway of 'purposive transition'. Geels and Schot (2007, p. 402) state that 'In our view, no transition is planned and coordinated "from the outset"'. Instead, they argue that every transition becomes coordinated at some point through the alignment of visions and activities of different groups. Nor does the model pay much attention to the issue of the speed of transition, emphasizing transition as 'a gradual, continuous process of change' (Rotmans et al. 2001, p. 16). In this connection, Victor et al. (2019) acknowledge that the nature of the LCT is quite different from historic technological transitions on at least three fronts: (1) it is problem-oriented (rather than opportunity-driven); (2) it cannot take decades as old ones used to; (3) incumbents are better organized and entrenched, thus better able to resist change. The implication of this combination is that policy makers must become crucial drivers of MLCT.

Numerous scholars have applied the governance perspective to studying China's LCT, often oblivious of the contradictory nature of the perspective and China's authoritarian context (i.e., a lack of strong non-state actors) (e.g., Xue et al. 2016; Yuan et al. 2012; Hu et al. 2015). Among the 115 articles reviewed by Wieczorek (2018), 11 are China-related. The review finds that studies of Chinese sustainability transitions have almost exclusively adopted the perspectives of the MLP and, to a much lesser extent, SNM. The article by Yuan et al. (2012) represents one of the earliest efforts to apply the MLP to the study of China's energy transition. They first characterized the present energy regime in China using the MLP. They then used the typology of transition pathways identified by Geels and Schot (2007) (reproduction, transformation, substitution, de-alignment/re-alignment, re-configuration) to describe potential transition pathways in China. They also made use of the TM framework and innovation theory to explore the management of the proposed transition pathways. In contrast, Andrews-Speed (2012) and Elliot and Zhang (2019) have used some forms of TM in their respective examinations of energy transition and the development of the green bonds market in China.

Combining new institutionalism with insights from the TM framework, Andrews-Speed (2012) has examined energy governance and its implication for the prospect of LCT in China. He explores institutional design at three different levels (culture, institution, rules governing individual interactions) and concludes that, while institutional adaption has taken place at level 3, these modifications and their efficacy have been constrained by level 2 institutions, that is, the formal institutional environment. Moreover, by placing emphasis on the embeddedness of the socio-technical regime around energy and the constraining nature of prevailing institutions, including an inactive civil society, Andrews-Speed (2012) stresses a lack of adaptive capacity by the Chinese system around energy and the LCT. He concludes that China's path towards LCT 'is likely to be very gradual' and to be closer to the 'business as usual' and 'reasonable effort' scenarios. China's subsequent MLCT trajectory suggests, however, that this prediction is significantly off the mark.

Elliot and Zhang (2019) show that the TM framework, once extended to consider transnational networks and learning, can be usefully applied to accounting for, at least partly, the development of the green bonds market in China: this development has been accompanied by growing interaction with widening transnational policy networks. Also acknowledging the role of international collaboration, however, Zhang (2019) stresses that the development of China's green bonds market is mainly driven by domestic factors, most importantly the financialization of the economy and the pursuit of an Ecological Civilization. Through the lens of interactive governance, Mai and Francesch-Huidobro (2014) analyse the governance of climate change mitigation responses in three major cities (Guangzhou, Shenzhen, and Hong Kong) in Southern China to explore how collaborative municipal networks functioned by focusing on two carbon intensive sectors, building and transport. They find that effective coordination relied on the political will of local administrative elites, the political significance attached to climate change issues, the legitimate authority granted to the coordinating agency, and human and financial capital. In a way, their study inadvertently underscores the importance of state actors in the MLCT processes.

To conclude, with its emphasis on non-state actors and horizontal relationships and its lack of attention to the scope and speed of the MLCT, the governance model is ill-suited for analysing China's MLCT experience. Having said that, it can act as a useful supplementary perspective in understanding this experience through its emphasis on the mobilization of multiple actors, networks, learning and experimentation.

Discussion

The discussions above examined how the three mainstream perspectives on MLCT (i.e. the policy, politics and governance models) have informed the existing understanding of China's MLCT experience. The key variables in these three models are respectively carbon pricing, horizontal power dynamics, and the role of non-state actors. In theory, the political model has the most potential if one follows Parsons' (Parsons 1963) framework of political power, which attaches importance to political leadership in setting collective goals, mobilizing resources, and ensuring compliance by other actors through fourfold strategies of inducement, coercion, persuasion and activation of commitments. However, such a view has few supporters and is shunned in transition studies. Moreover, Parsons' framework is highly abstract and is lacking in the explanation of mechanisms by which goals are set, resources are mobilized, and compliance is secured (Giddens 1968). When applied to China, although extant studies using these models have highlighted the importance of the government's prioritization and policies to low-carbon development in discrete areas, they have not adequately explored the overall motivation, strategies, and practices of state actors. In cases where they do, researchers are often swayed by the FA model. Such a combination has resulted in a greater focus on what has failed to work than on what has worked. On the other hand, with few exceptions (e.g., Lo 2014), there is a paucity of systematic effort to examine the relationship between governing arrangements and their MLCT outcome.

MY PERSEPCTIVE: GOVERNMENTALITY

For the reasons stated above, we have come to think that it is necessary to change the lens of analysis to gain a better understanding of the Chinese experience. The new lens that we have chosen is governmentality. It is essentially 'a strategical model' of power in Foucault's words, drawing attention to objectives, tactical efficacy, and the analysis of a multiple and mobile field of force relations (Foucault 1978, p. 102). Taking up this perspective implies a search for a carbon governmentality (or governmentalities) underpinning the guidance of the conduct of those groups that are crucial to MLCT. Extending the introduction in Chapter 1, we seek to explain this lens and operationalize it below.

Defining Key Terms

The perspective of governmentality originates from the work of Michel Foucault. However, it has come to encompass contributions from others. Foucault used the terms 'rationality of government' and 'governmentality' almost interchangeably with 'the art of government'. By 'rationality of government', he refers to 'a way or system of thinking about the nature of the practice of government (who can govern; what governing is; what or who is governed), capable of making some form of that activity thinkable and practicable' (Gordon 1991, p. 3). The term 'rational' here 'refers to the attempt to bring *any* form of rationality to the calculation about how to govern' (Rose 1999, p. 18). Foucault employs three general forms of rationality: the concepts of strategies, technologies, and programmes of power. These are related respectively to certain forms of explicit and reflected discourse, certain non-discursive social and institutional practices, and certain effects produced within the social field (Gordon 1980, p. 246).

It is necessary to recognize that Foucault's rationality is at the level of the system, rather than associated directly with any individual or organization. Foucault (1978) states: 'the rationality of power is characterized by tactics that are often quite explicit at the restricted level where they are inscribed (the local cynicism of power), tactics which … end by forming comprehensive systems: the logic is perfectly clear, the aims decipherable, and yet it is often the case that no one is there to have invented them, and few who can be said to have formulated them' (p. 93). The Foucauldian study of rationality seeks to identify the overall logic of the system that emerges from power relations. It aims to discover a strategy that invests the regimes of practices 'with a relative unity and functionality in relation to a dynamic and variable set of objectives' (Dean 1996, p. 61).

Historically, Foucault (1991a) identifies three characteristic kinds of governmentality, associated with three distinctive forms of political power. The latter are: (1) *the state of justice*, born in the feudal type of territorial regimes corresponding to a society of laws; (2) *the administrative state*, born in the territoriality of national boundaries and corresponding to a society of regulation and discipline; and (3) *the governmental state*, 'which bears essentially on population and both refers itself to and makes use of the instrumentation of economic *savoir*', and corresponds to a society controlled by the apparatus of security (Foucault 1991a, p. 104). The third type, prevalent since the mid-eighteenth century, has two distinctive features: its reliance on a body of knowledge – political economy; and its purpose as 'not the act of government itself, but the welfare of the population, the improvement of its condition, the increase of its wealth, longevity, health, etc.' (Foucault 1991a, p. 100). He calls the third type of power 'biopolitics' due to its focus on population and the use of knowledge about the human body and soul. Foucault (1991a) points out that what we have today is a triangle, sovereignty-discipline-government (p. 102). In other words, contemporary nation states deploy a combination of juridical, disciplinary and government powers. The implication of the triangle is that we need to analyse contemporary government practices as an aggregate of the exercise of all three kinds of power. It also raises the question of how to combine these powers.

Theoretically, Foucault (1991a) uses the term governmentality to refer to three different things in the context of liberal democracies. First, he uses it to refer primarily to the 'ensemble formed by the institutions, procedures, analyses and reflections, the calculations and tactics' that allow the exercise of government power (p. 102). Second, it refers to the historic tendency towards the pre-eminence of government power over powers of sovereignty and discipline, resulting in the formation of a whole series of governmental apparatus and a complex of '*savoirs*'. Third, he uses governmentality to refer to the governmentalization of the state, 'the process by which the institutions of the state become transformed to support a governmentality based on biopower' (Oels 2005, p.189). In this sense, a study of carbon governmentality is about identifying the ensemble of institutions, procedures, analyses and reflections, the calculations and tactics that make it possible for the Chinese state to curb carbon emissions and accelerate LCT while paying attention to laws and regulations.

Why the Governmentality Perspective?

The governmentality perspective has several features that make it useful for the analysis of Chinese MLCT as a target of government. These include: a unique conception of power including a central role for knowledge and discourse;

a positive view of power; a focus on the question of 'how' to govern; insights into 'governing at a distance' and the role of translation; a distinctive approach to history; and liberalism and authoritarianism as distinct but connected forms of governmentality. These are further discussed below.

Foucault has introduced a new way of conceptualizing and analysing political power by power through the concept of governmentality. First of all, Foucault (1978) does not mean a group of institutions and mechanisms that would ensure the subservience of the citizens, or a mode of subjugation, or a general system of domination exerted by one group over another. Rather he defines power foremost as 'the multiplicity of force relations immanent in the sphere in which they operate and which constitute their own organization', but also as the process that transforms, strengthens or reverses the relations, as the support that these relations find in one another, and as 'the strategies in which these relations take effect' and become embodied in state apparatus, laws and various social hegemonies (Foucault 1978, pp. 92–3). This means that there is no substantiation of power, only power relations. It also implies that the state is not the ultimate source of power. Rather it must work with other forces, especially those from below, by deploying a range of governmental strategies, technologies and programmes.

Second, Foucault places what he calls 'power-knowledge relations' and discourse at the heart of governmentality. In *Discipline and Punish*, he writes: 'We should admit rather that power produces knowledge (and not simply by encouraging it because it serves power or by applying it because it is useful); that power and knowledge directly imply one another; that there is no power relation without the correlative constitution of a field of knowledge, nor any knowledge that does not presuppose and constitute at the same time power relations' (Foucault 1977, p. 27). He shows, for example, that the power to punish derives its bases, justification and rules from the knowledge about the criminal and the mastery of the force relations that he is involved in. On the other hand, Foucault (1978) places discourse at the heart of power-knowledge, saying that 'it is in discourse that power and knowledge are joined together' (p. 100). He defines discourse as 'a thing pronounced or written' (Foucault 1970, p. 52). He argues that 'discourse is not simply that which translates struggles or systems of domination, but *is the thing for which and by which there is struggle, discourse is the power which is to be seized*' (Foucault 1970, pp. 52–3, my emphasis). Moreover, in Foucault's analysis, discourses are both dynamic and tactical: they are 'a series of discontinuous segments whose tactical function is neither uniform nor stable'; they are 'tactical elements or blocks operating in the field of force relations' (Foucault 1978, p. 100).

Third and related to the previous point, Foucault conceptualizes power as a productive force. Foucault (1978) criticizes the traditional negative conception of state power, dismissing it as putting too much stress on prohibition and rules. He points to the possibility of an alternative conceptualization of state power that 'exerts a positive influence on life, that endeavors to administer, optimize, and multiply it, subjecting it to precise controls and comprehensive regulations' (p. 137). Foucault (1984) states: 'what makes power hold good ... is simply the fact that it does not only weigh in on us as a force to say no, but that *it traverses and produces things*, it *induces pleasure, forms knowledge, produces discourse*. It needs to be considered as a productive network which runs through the whole social body, much more than as a negative instance whose function is repression' (p. 61, my emphasis). Power/knowledge is precisely what underpins governmentality and ultimately enables governmental power to achieve its objectives. In the words of Oels (2005, p. 186) 'power is productive by constituting subjects and objects and by inciting discourses'.

The governmentality perspective is uniquely focused on the question of 'how' to govern. Thus, analytics of government involves more than an examination of the administrative structure, organizations and personnel involved and their expertise, the means of managing information, procedures, and the like. Instead, it attends to how all the above is thought and formed in relation to specific forms of knowledge and expertise of various authorities, and finally how certain forms of thought such as specific policies or programmes seek to unify and rationalize the techniques and practices relative to objectives and problematizations, and evaluative schemes (Dean 2010).

Governmentality also sheds a unique light on 'governing at a distance' and the role of translation in this endeavour. This is particularly relevant to a large country like China. Although this aspect was not emphasized in Foucault's original work, it has emerged from scholarly efforts to link the calculations, strategies and programmes of the centre to activities far distant in space and time (e.g., empires) and to events in thousands of settings through processes of 'translation'. Rose (1999, p. 48, my emphasis) argues:

> Clearly a plan, policy or programme is not merely 'realized' in each of these locales, nor is it a matter of an order issued centrally being executed locally. What is involved here is something more complex. I term this 'translation'. *In the dynamics of translation, alignments are forged between the objectives of authorities wishing to govern and the personal projects of those organizations, groups and individuals who are the subjects of government.*

Rose (1999, p. 50) also explains that translation 'involves processes which link up the concerns elaborated within rather general and wide-ranging political rationalities with *specific* programmes for government of this or that problematic zone of life'. He acknowledges that translation is an imperfect mechanism and is subject to pressures and distortions. In this light, the so-called 'fragmented authoritarianism' in China is a normal, rather than a pathological state.

The governmentality perspective is complemented by a distinctive approach to the history of ideas, values and practices, described as the genealogical approach. Foucault (1984, p. 59) defines genealogy as 'a form of history which can account for the constitution of knowledges, discourses, domains of objects, etc.' without having to relate them to a transcendental or unyielding subject. Foucault points out that words may change their meaning; desires may point in different directions; ideas may assume different logics. Foucault (1978) was critical of traditional history in its search either for a single origin or finality, such as utility or rationalization. Rather the genealogical studies 'record the singularity of events' and 'seek them in the most unpromising places' such as sentiments, love, conscience and instincts. They provide scope to explore the links between discourses of the present and the past, and between discourses in different domains at the same points in time. The genealogical approach alerts us to the need to identify and examine how specific situations may have caused an old art of governing to be called into question and new ones to come into being.

Finally, the perspective of governmentality suggests that the distinction between liberalism and authoritarianism is not as stark as it is often made out to be. Dean (2010) argues that, although governmentality is developed in the context of liberal democracies, it could be applied to thinking about the rationalities and technologies of authoritarian forms of rule in cases such as contemporary China. He explains that '"authoritarian governmentality", like liberal and social forms of rule, is made up of elements assembled from bio-politics and sovereignty. Further ... authoritarian governmentality can also be located along the trajectory of the governmentalization of the state' (Dean 2010, p. 155). In this perspective, authoritarian rule is simply characterized by more elements of sovereignty power, and presumably disciplinary power, than the biopolitical one. This understanding makes it more fathomable to discuss lessons that can be learnt from China by other nation states. Nevertheless, there is a qualitative difference between liberalism and non-liberalism. Liberalism is marked by a respect for the natural courses of things and a new rationality in the art of government: 'governing less, out of concern for maximum effectiveness, in accordance with the naturalness of the phenomena one is dealing with' (Senellart 2008, p. 327). This raises the question about the extent to which authoritarian rules are capable of accommodating, and indeed utilizing, elements of liberal rationality in the pursuit of maximum effectiveness.

Governmentality as Applied to Climate/Carbon Governance

Some efforts have been made to apply the governmentality perspective to studies of carbon governance internationally (e.g., Okereke et al. 2009; Bulkeley et al. 2014). Bulkeley et al. (2014), for example, use governmentality to frame forms of experimentation in climate change interventions as part of the 'art of government'. They argue that 'the governing of climate change is orchestrated by the competing rationalities, techniques and tactics that actors mobilize as they seek to define both the object to be governed and the subjects with whom it is concerned' (Bulkeley et al. 2014, p. 33). Another approach emphasizes the potential of governing strategies and the positive loop between carbon government and the strengthening of state capacity. Rutland and Aylett (2008) stress the strategies that modern governments can employ to guide the conduct of citizens and achieve government objectives. They argue that 'when individuals come to view themselves and their goals according to the same metrics as the state, and base their actions on these metrics, they become part of the network of self-regulating actors that is at the heart of the practice of governmentality' (Rutland and Aylett 2008, p. 631). On the other hand, with the example of Seattle in Washington, Rice (2010) addresses the specific question of why local authorities may wish to engage in carbon control without mandates from upper governments. She suggests that carbon governance enables the local government to reproduce governing capacities and reaffirm its regulative ability in areas ranging from infrastructural design to commercial activities and neighbourhood development. She also makes the interesting point that carbon control makes climate change relevant to the practice of the state because carbon marks the territorial logic of the state and because of its measurability and quantifiability.

Furthermore, Wieczorek (2018) identifies several articles adopting Foucault's positive and relational conception of power. The article by Tyfield (2014) is particularly relevant, as it relates to China and explores the key contributions that a Foucauldian-inspired cultural political economy can offer in filling some of the gaps in applying the MLP framework to studies of mobility in China. Tyfield criticizes the MLP's negative overtone regarding power and politics, where regime actors, including government officials, tend to be regarded as resistant to change. He suggests that placing the conception of productive power at the heart of the analysis enables us to see the emergence of the new socio-technical regime as a power transition: 'this process is the outcome of an essentially dynamic process of multiple strategic agencies jostling for position and the connections and coalitions they actively seek in this pursuit' (Tyfield 2014, p. 591). He argues that, in this perspective, existing efforts to bring power and politics to the MLP miss the point, as the focus in these efforts has been on explaining the stability and 'lock-in' of existing systems and the exploitation of non-discursive 'power' by incumbent actors. Instead, he recommends that we should 'begin to think about the crucial question of the interaction between specific socio-technical systems and broader power regimes' (Tyfield 2014, p. 592).

In operationalizing the governmentality perspective for climate action, the articles by Oels (2005) and Paterson and Stripple (2010) are instructive. Both have used Dean's 'analytics of government' framework. Oels (2005) explores how the changing role and form of government has influenced discourse on climate change internationally. She argues that 'the ways in which climate change is rendered a governable entity are best understood before the background of a shift from biopower to advanced liberal government' (p. 185). On the other hand, drawing on Dean (1999) and Oels (2005), Paterson and Stripple (2010) highlight the importance of moulding and mobilizing people's individual subjectivity and capacities to govern themselves for addressing climate change. They apply the lens of the analytics to studying five kinds of calculative practices by individuals in managing their own emissions. They conceptualize these practices as the behaviour of carbon conduct and characterize this conduct as 'government enabled through certain forms of knowledge (measurements and calculations of one's own carbon footprint), certain technologies (the turning of carbon emissions into tradable commodities), and a certain ethic (low-carbon lifestyle as desirable)' (Paterson and Stripple 2010, p. 347). These studies have highlighted the importance of understanding the specific climate governmentality in the context of general governmental shifts. They also underline the need to identify the forms of knowledge, technologies and ethics utilized in climate governmentality.

Governmentality in China

The concept of governmentality has been developed in the context of Western countries. Yet the governmentality perspective has been applied to studying a variety of topics in China, ranging from the *Danwei* (work unit) system (Bray 2005) to the one-child policy (Sigley 1996, 2006), population governance (Greenhalgh and Winckler 2005), and urban community life (Wan 2016). Sigley (1996) made one of the earliest attempts to apply the perspective of governmentality to the Chinese context. He highlights three enduring characteristics of the Chinese government's approach. First, there is a high level of confidence and control by the government. The 'plan' is characterized as 'a masterpiece of total calculation' (Sigley 1996, p. 494). Moreover, the government 'argues that through the sciences of Marxism-Leninism there is nothing it does not know' (Sigley 1996, p. 494). Second, rather than requiring the expert intervention of professionals in implementing their plans, as governmentalities elsewhere tend to do, Chinese governmentality relies heavily on SNGs. In the case of controlling population growth through the one-child policy, for instance, this system consisted of the determination of birth quotas, the distribution of such quotas across provinces/prefectures and the stipulation of incentives and disincentives by the centre on the one hand, and the translation of these quotas into permitted number of pregnancies and monthly monitoring cycles at local level on the other hand. Third and finally, Sigley (2006, p. 491) characterizes Chinese governmentality as governing 'not through familiar tactics of "freedom and liberty", but through a distinct planning and administrative rationality'.

Meanwhile, researchers of Chinese governmentalities have highlighted significant shifts in governmentality in China because of the post-Mao economic reforms. Sigley (2006) suggests that the introduction, from the early 1990s, of the 'socialist market economy' (SME) requires the state to intervene in different ways by combining neoliberal and socialist strategies. This has led to a new hybrid form of socialist-neoliberal political rationality that is at once 'authoritarian in the familiar political and technocratic sense' and seeks to govern selected subjects 'through their own autonomy', in terms of market mechanisms for example. Another shift that he observes is the redeployment of a long-held 'realist' or 'social Darwinian' view of the world in terms of competing nation states. Sigley (2006, p. 496) also notes the changing functions of government, including 'the shift in vocabulary and conceptualization within Chinese discourse from a notion of "government" (zhengfu) as a task of "planning" (jihua) and "administration" (xingzheng) to one that involves "management" (guanli) and "governance" (zhili)'. Third and finally, Chinese governmentality is becoming less fixated on economic growth. Greenhalgh (2005, p. 25) notes that in response to the challenges of a changing world economy, in the twenty-first century, China's biopolitics of population 'is mutating, becoming less econometric and more focused on social and even human governance'. In a rare example of using governmentality to study the ecological movement in China, Wang and Zhang (2019) examine the emergence of the discourse around eco-cities. They suggest that this shift towards eco-cities is 'a state-initiated and framed response to development challenges faced by Chinese cities' (p. 38). However, their focus is more on the 'eco-city mentality' than on the practices of realizing the eco-city vision.

ANALYTICS OF CARBON GOVERNMENT: OPERATIONALIZATION

Rose and Miller (1992, p. 175) suggest that governmentality may be analysed at two levels: political rationalities and their governmental technologies. While the former are defined as 'the changing discursive fields within which the exercise of power is conceptualised, the moral justifications for particular ways of exercising power ... notions of appropriate forms, objects and limits of politics, and conceptions of the proper distribution of such tasks', the latter are defined as 'the complex of mundane programmes, calculations, techniques, apparatuses, documents and procedures through which authorities seek to embody and give effect to governmental ambitions' (Rose and Miller 1992, p. 175). Describing the technologies of government as 'human technologies', Rose (1999, p. 52) states that 'within these assemblages, it is human capacities that are to be understood and acted upon by technical means'. More concretely, Dean (2010) has proposed an '*analytics of government*' for studying governmentality. He defines this as a process that 'examines the conditions under which regimes of practices come into being, are maintained and are transformed' (p. 33), characterizing such regimes as 'fairly coherent sets of ways of going about doing things' (p. 31). He further explains that an analytics of government will 'seek to constitute the intrinsic logic or *strategy* of a regime of practices that cannot be simply read off particular programmes, theories and policies of reform' (p. 32). Instead, the strategic logic can only be constructed through understanding its operation as an intentional but non-subjective assemblage of all its elements and is irreducible to the explicit intention of any single actor, since it is the result of power interaction.

Dean (2010, pp. 33–44) identifies four different, reciprocally conditioning, yet relatively autonomous dimensions of this analytics:

1. Characteristic forms of visibility, ways of seeing and perceiving.
2. Mentalities or distinctive ways of thinking and questioning (derived from sciences and informing the practices).
3. Specific ways of acting, intervening and directing.
4. Characteristic ways of forming subjects, selves, persons, actors or agents.

He makes it clear that this analytics is not designed to airbrush blemishes of the system. Instead, he suggests that a source of critical potential of the analytics of government 'comes from showing the "inconvenient" dissonance between the claims and objectives of programmes and rationalities of government and the "intentional but non-subjective" character of regimes of practices, that is to say, their logic, their intelligibility, and even their strategy' (Dean 2010, p. 4). Dean's four-dimensional framework has been helpfully turned into a matrix by Oels (2005), with a characteristic set of research questions. Building on these foundations, below I seek to operationalize the analytics of carbon government in China by clarifying the four dimensions. Following Rose and Miller's (1992) twofold distinction between political rationalities and governmental technologies, we will place Dean's non-technical aspects, including the field of visibility, forms of knowledge, and formation of subjects, under 'political rationalities'.

Field of Visibility

This is about visualizing the field to be governed (Dean 2010). The purpose is to make it possible to picture 'who and what is to be governed, how relations of authority and obedience are constituted in space, how different locales and agents are to be connected with one another, what problems are to be solved and what objectives are to be sought' (Dean 2010, p. 41). Using the governmentality perspective brings into focus several key objects: levels, sources and distribution of GHG emissions; categories of GHG emitters; those who are in a position of authority to affect emissions and related behaviours; the official mitigation pledges and planning objectives, and the high carbon intensity of the economy as the problem to be resolved. Our analysis will show that what distinguishes the Chinese system is the dual roles of SNGs, both as the governed and the governor, and the attention paid to strengthening their roles and capacities in managing MLCT.

The Approach to Government as a Rational and Thoughtful Activity

This dimension concerns the forms of knowledge that arise from and inform government. Dean calls it '*the epistēme* of government' as the domain of veridical discourse and theoretical knowledge (Dean 1996). It is manifested in policy programmes. Key questions here include: 'what forms of thought, knowledge, expertise, strategies, means or calculation, or rationalities are employed in practice of governing? How does thought seek to transform these practices? How do these practices of governing give rise to specific forms of truth? How does thought seek to render particular issues, domains and problems governable?' (Dean 2010, p. 42).

The Foucauldian analysis of discourses puts emphasis on their production function, their role in the constitution of power relations, affecting a strategic situation, and in the construction of subjects (Feindt and Oels 2005). The term 'subject' here points to an actor in possession of his or her own identity by a conscience or self-knowledge and being subjected to someone else by control and dependence at the same time (Foucault 1982, p. 781). In the context of discourses on sexuality, Foucault (1978, p. 102) once said that 'we must question them on the two levels of their tactical productivity (what reciprocal effects of power and knowledge they ensure) and their strategical integration (what conjunction and what force relationship make their utilization necessary in a given episode of the various confrontations that occur)'. Feindt and Oels (2005, p. 165) suggest that 'A Foucaultian discourse analysis of environmental policy making would have to show how political problems and a related set of subjects and objects are discursively produced and rendered governable.' Through the governmentality lens, we will explore how changing discourses, learning, and knowledge creation have enabled the problematization of the existing practices and the production of subjects and objects for carbon government in China.

Identities and Identification

This dimension is concerned with 'forms of individual and collective identity through which governing operates and which specific practices and programmes of government try to form' (Dean 2010, p. 43). Key questions here relate to forms of person, self and identity that are presupposed by the regime of practices of government and the sorts of transformation that are sought; status, capacities, attributes and orientations that are assumed of those who exercise authority and those who are to be governed; forms of conduct that are expected of them; duties and rights that they have; ways in which these capacities and attributes are to be fostered; and ways in which these duties enforced and rights ensured. Dean stresses two related points. On the one hand, the forms of identity promoted and supposed 'should not be confused with a *real* subject … i.e. with a subject that is the endpoint or terminal of these practices and constituted through them' (Dean 2010, p. 43). Rather regimes of government only 'elicit, promote, facilitate, foster and attribute various capacities, qualities and statuses to particular agents' (p. 44). They are successful to the extent that these agents come to experience these capacities, qualities and statuses. A key mechanism here is the self-conceptualizations of the agents, constructed in a field of discourses (Ettlinger 2011). On the other hand, much of the problem of government here 'is less one of identity than one of "identification"' according to Dean (2010, p. 44). This refers to the idea that in postmodernity, experiencing something together has assumed greater significance than in modernity and works as a cement for society (Maffesoli 1991). It points to the importance of appealing to collective emotions and exploiting aesthetics including sensibility, sensation, sentiment and attraction for the purpose of constructing responsible carbon managers. The formation and transformation of identities and the mobilization of identification could well be challenging for an authoritarian regime.

Technical Aspects

The technical aspects refer to the domain concerned with forms of practical knowledge. A useful distinction is between 'means of government' and 'technologies of government'. While the former refer to 'humble and mundane mechanisms' such as techniques of notation, computation and calculation, procedures of examination and assessment, new vocabularies and architectural designs (Miller and Rose 1990), the latter are much more strongly oriented towards the conduct of conduct and the optimization of performance. Dean (1996, p. 48) argues that 'technologies of government are distinguished by a particular orientation toward conduct that takes the form of a strategic rationality concerned with the optimization of performance, aptitudes and states'. This is achieved by making aspects of conduct calculable, measurable and comparable.

Dean (1996) marks out four threshold conditions for making a system of government technological. The first condition refers to the existence of a complex assemblage of diverse elements, held together by heteromorphic relations and all concerned with the direction of conduct. It is characterized by 'explicit forms of practical rationality, such as programmes, policies and plans, that invest this technology with a certain set of purposes and codify its functioning, and by specialist knowledge or experience that acts as a means of calculation and allows planning and evaluation to occur' (Dear 1996, p. 64). Second, it is attached to other kinds of technological systems. Third, 'the forces and capacity made available are qualitatively different from any simple argumentation or synthesis of existing forces'; they make possible the constitution of specific sites as 'power containers', 'power storers' and 'power-generators' and a coordination of activities across these different locales in time and space. Thus, the powers they constitute are 'infrastructural powers' and exhibit a level of durability. Fourth and most importantly, it is 'invested with a strategic rationality that seeks to subsume the moral and political shaping of conduct to the requirements of performance' (p. 63).

In other words, what makes a system of government technological is the existence of an identifiable assemblage, interlinkage between different kinds of technological systems, the creation of new durable power sites carrying infrastructural powers, and a strong orientation towards performance and aptitude optimization. In this perspective, attention ought to be paid to different forms of practical rationalities as well as the institutional arrangements and institutional sites that store and generate power, especially infrastructural powers.

The operationalization of the analytics of China's carbon government is summarized in Table 2.1. We will explore the non-technical and technical aspects in Chapter 3 and Chapter 4, respectively.

Table 2.1   *Analytical framework for studying carbon governmentalities in China*

	Analytical category	Key questions	Most relevant parts of this book
Political rationalities	Field of visibility	What is illuminated?	Chapters 3, 5, 6
		Problems to be solved	
		Objectives to be sought	
	Forms of knowledge	What forms of thought and knowledge arise from and inform the activity of governing?	Chapters 3, 4, 5, 6
	Formation of identities	What forms of subject, self, person, actor or agent are presupposed by practices of government?	Chapters 3, 5, 6
		Which transformations are sought?	Chapters 4, 5, 6
Governing techniques and technologies	Technical aspects	By what means, mechanisms, procedures, instruments, procedures, tactics, techniques, technologies, and vocabularies is authority constituted and rule accomplished?	Chapters 4, 5, 6

Source: Adapted from Rose and Miller (1992), Dean (2010, pp. 41–4) and Oels (2005).

3. Beneath China's low-carbon transitions: political rationalities

INTRODUCTION

One of Foucault's greatest insights concerns the pivotal role of reasoning and strategies in the exercise of power. To make sense of China's political rationalities for the active pursuit of mitigation and low-carbon transitions (MLCT), it is necessary to examine the field of visibility including the problems to be solved and objectives to be achieved, the forms of knowledge that have informed and arisen from the activity of governing, and the forms of identity presupposed and their transformation sought (see Table 2.1). Lemke (2002, p. 55) reminds us that the problem for a Foucauldian analysis is not to investigate if practices conform to rationalities but to discover which kind of rationality they are using. This chapter examines how different kinds of rationality as discourses, strategies and programmes have emerged and transformed in the Chinese context, and how such shifts have affected the constitution of governing objects and subjects for MLCT, thus prefiguring the technical aspects of China's carbon governmentality. The latter will be examined in Chapter 4.

This chapter is organized into two main parts. The first part starts with a sketch of the broad discursive field in China and its shifts since 1949, when the People's Republic of China (PRC) was founded. This long-lens genealogical review is in recognition of the fact that discourses often have deep and convoluted roots. The second part then examines the three dimensions of political rationalities specific to carbon governmentality in China.

IDEOLOGY AND SHIFTING DISCOURSES IN CHINESE POLITICS

Discourse plays an extremely important role in Chinese politics and governance. It was Marxist discourse that gave birth to the Chinese Communist Party (CCP) in 1921 and guided its ultimately victorious struggle to gain power in 1949. Moreover, ideology (*yi shi xing tai* in Chinese) continues to play a hegemonic role in the sense that, through carefully formulated official discourses, the party seeks to legitimize its one-party rule by adopting policy goals that have support and consent from the Chinese society. This is because hegemony 'is especially important in societies in which electoral politics are not established and authoritarian governance prevails' (Su 2011, p. 312). On the other hand, consistent with the theses of Hall (1993) and Schmidt (2008), economic policy making in China is deeply influenced by economic paradigms that the elites adopt (Brødsgaard and Rutten 2017). Without considering the reciprocal dynamics between discourse, politics and economic outcomes, Brødsgaard and Rutten (2017) argue, it is difficult to understand why the Chinese economic system has developed by way of a periodic succession of distinct and sometimes contradictory modes of governance. In their view, such successions result from unforeseen imbalances in the economy that then necessitate a re-conceptualization of economic paradigms in relation to the issues of accumulation, readjustment and reform.

A fundamental feature of post-Mao Chinese politics is that the party-state claims its legitimacy based not only on socialism as a progressive ideology, but more importantly on its capacity and track record in promoting the modernization of the country and improving the standards of living for the large population. In other words, the system depends on 'performance legitimacy' (or *zhengji* in Chinese) (Wang 2013). Deng Xiaoping, the architect of China's post-Mao economic reform, put this logic most succinctly:

> If we did not adhere to socialism, implement the policies of reform and opening to the outside world, develop the economy and raise living standards, we would find ourselves in a blind alley. We should adhere to the basic line for a hundred years, with no vacillation. That is the only way to win the trust and support of the people. (Deng 1993, pp. 370–1)[1]

In this connection, Dirlik and Zhang (1997, p. 8) attribute the success of the post-Mao Chinese system to 'being more successful at capitalism than capitalist societies'. However, some Chinese policy makers have recognized the shortcomings, or even danger, of a pure economic performance logic (e.g., Pan 2004). Instead, top CCP leaders have attempted to continuously reinvent the system by broadening its appeal. While the official discourses in the post-Deng era continue to put emphasis on economic development and modernization, there have been two important shifts. These involve respectively the incorporation of an environmental/ecological rationality from the late 1990s to the early 2000s and the development of a form of internationalism from the

---

[1]   The English translation here is based on 'Excerpts from Talks Given in Wuchang, Shenzhen, Zhuhai and Shanghai', Deng Xiaoping (18 January–21 February, 1992), https://olemiss.edu/courses/pol324/dengxp92.htm (accessed 4 July 2020).

mid-2010s. We argue that these shifts have facilitated the emergence of carbon governmentality in China.

Modernization as a National Goal After 1954

The CCP has consistently presented itself as a modernizing force in China. The PRC's first constitution (NPC 1954) argues that the CCP led the Chinese people to victory in their historic struggle against the suppression and enslavement imposed upon them by imperialism, feudalism and bureaucratic-capitalism, and that socialism under the leadership of the CCP is an assured way to 'banish exploitation and poverty and build a prosperous and happy socialist society' through socialist industrialization and transformation. When the CCP-led government issued its first Government Work Report in 1954, it proposed the modernization of industry, agriculture, defence, transport and communication as a national goal (Zhou 1954). This was the original 'four modernizations' programme. The current Four Modernizations programme, encompassing the modernization of industry, agriculture, defence, science and technology, was first adopted in 1964 at the Third National People's Congress, the highest legislative body, and marks a heightened emphasis on science and technology in comparison with the original 'four modernizations'. It was envisaged then that the modernization process would take two steps: first, to establish an independent and relatively comprehensive industrial system and national economy system by 1980; and second, to realize comprehensively the four modernizations by the end of the twentieth century (Zhou 1975). This plan proved to be over-optimistic, as the ideologically driven ten-year Great Cultural Revolution (1966–76), which focused on 'class struggle', disrupted the implementation of the 1964 plan. After Mao's passing, to move away from 'class struggle', the paramount leader Deng Xiaoping renewed the commitment to the Four Modernizations in 1975 (Peng 2020).

Economic Reform and the Shift towards Economic Development

After the removal of the so-called 'Gang of Four' in 1976, the new leadership faced a daunting task of rebuilding the CCP's legitimacy amidst widespread poverty and social chaos. The strategy adopted by Deng Xiaoping and his supporters was to introduce a reform policy, including opening up, to realize the Four Modernizations. The decision of the historic Third Plenum of the 11th CCP Congress states: 'The focus of the whole party's work from 1979 should shift to the socialist modernization drive.' It further acknowledges that 'to realize Four Modernizations requires a substantial increase in productivity.' At the 13th Party Congress in 1987, the Four Modernizations programme was given specific formal milestones by Zhao Ziyang, the then Secretary General of the CCP Central Committee (CCPCC), drawing on Deng's ideas back in 1984. Characterized as a 'three-step strategy' towards China's modernization, these milestones include: (1) solving the problem of inadequate food and clothing by 1990; (2) realizing a comfortable living (*xiaokang* in Chinese) by 2000; (3) achieving the Four Modernizations by the middle of the twenty-first century (Zhao 1987). This strategy has been followed through by all subsequent leaders, although its specific wordings have changed over time. The current formulation was first set out in a speech to the 15th CCP Congress in September 1997 by the then General Secretary of the CCPCC, Jiang Zemin. He declared:

> Looking into the next century, we have set our goals as follows. In the first decade, the gross national product will double that of the year 2000, the people will enjoy an even more comfortable life and a more or less ideal socialist market economy will have come into being. With the efforts to be made in another decade when the Party celebrates its centenary, the national economy will be more developed and the various systems will be further improved. By the middle of the next century when the People's Republic celebrates its centenary, the modernization program will have been accomplished by and large and China will have become a prosperous, strong, democratic and culturally advanced socialist country. (Jiang 1997)

Jiang's formulation has two notable new features. On the one hand, it identifies three medium- and long-term goals, corresponding with three milestones (i.e., 2010, 2020 and 2050). The second and third goals are called the 'Two Centenary Goals', referencing the foundation of the CCP and the PRC respectively. On the other hand, this formulation broadens the goal of the modernization programme to include not just a more developed economy, but also democratization and cultural advancement. In his report to the 16th Party Congress in 2002, Chairman Jiang rephrased the first centenary goal as 'building a well-off society in an all-round way'. Chairman Hu's (2012) report at the 18th CCP Congress reaffirmed commitment to the 'Two Centenary Goals'. Although the report by Chairman Xi at the 19th Party Congress added what is popularly described as the 'Chinese dream' of national rejuvenation (Xi 2017a), Xi (2014) explained in a speech in Berlin that the 'Chinese dream' is the same as the second Centenary Goal (i.e., completion of modernization). Clearly, China's modernization programme is much more than what the 'Four Modernizations' literally refer to.

The Debate on the Criterion of the Truth and the Discourse of 'Four Cardinal Principles'

In its drive to speed up modernization and develop the country's economy, the CCP has orchestrated a series of discourses over the past decades. One of the most significant took place early on, in May 1978, when Hu Yaobang, a vice-president of the Central Party School at the time and a future general secretary of the CCPCC (1982–7), launched a nation-wide debate on 'What is the sole criterion of truth?' through the CCP's key mouthpieces including the *Theoretical Trends, Guangming Daily* and *The People's Daily*. From Schoenhals' (1991) fascinating account of this episode, two key points can be drawn. First, this debate initiated by Deng and his supporters was designed to challenge the 'two whatevers' which were perceived as preventing the new leadership from breaking new ground by tying the present leaders to decisions made by Chairman Mao, the now deceased great leader.[2] Second, it elevated 'practice' or 'results in practice' to the status of being a sole provider of legitimacy. Enshrined in the CCP constitution, this principle confirms the effectiveness of promoting modernization as a paramount criterion for judging the work of any individual, organization or policy in China. Furthermore, coupled with the shift of the party's work towards economic development, this principle also lends support to the 'criterion of productive forces', thus giving 'an unprecedented degree of "truth" to rationalizations based on economic pro- ductivity' (Schoenhals 1991, pp. 267–8). This implies that any policy practice must prove its worth in promoting productivity. Foucault (2008, p. 35) pointed out that the art (or rationality) of government involves a regime of verification not by law, but by 'the set of rules enabling one to establish which statements in a given discourse can be described as true or false'. In this light, the CCP leadership has established a set of verification rules in stark productivity terms.

However, Deng was also responsible for introducing the so-called 'Four Cardinal Principles' (FCPs in short). These were first articulated in a speech by him in 1979 on behalf of the CCPCC, only months after the historic Third Plenum of the 11th CCP Congress, which established the primacy of economic development. These principles refer to: keeping to the path of socialism, upholding the people's democratic dictatorship, upholding the leadership of the Chinese Communist Party, and upholding Marxism-Leninism-Mao Zedong Thought.[3] They have been enshrined in the Chinese constitution since 1982 and have been identified as an important element of the party's 'basic line' since 1987. Deng's argument is that these principles are the prerequisite for realizing the Four Modernizations. The combination of these two dis- courses is striking. If the 'criterion of truth' discourse represents an attempt at government power, the FCPs are undoubtedly juridical in nature.

---

[2]   The 'two whatevers' refer to the pledge to 'resolutely defend whatever policy decisions Chairman Mao made, [and] steadfastly abide by whatever instructions Chairman Mao gave' (Schoenhals 1991, p. 249).
[3]   This English rendering is based on China Translation Institute Office (http:// keywords.china.org.cn/2018-10/30/content_69101990.htm) (accessed 18 July 2019).

Socialist Market Economy

Following the so-called 'Tiananmen Incident' in 1989, where many pro-democracy protesters were reportedly killed by the army under the order of the top leaders including Deng Xiaoping, the party-state went into a period of reflection and retrenchment about its reform policy. While those on the left called for political democratization, those on the right advocated retreat to the planned economy and a slow-down of the reform programme. Questions were raised about the socialist or capitalist genealogy of some of the reform measures. This prompted Deng to undertake his famous 'southern inspections' in early 1992, during which he inspected a number of places. By Deng's (1993) own account, he was tremendously encouraged by what he saw, especially in the two Special Economic Zones of Shenzhen and Zhuhai, in terms of the progress result from the reform and opening policy. Subsequently, he advocated furthering rather than retreating from reform. Referring to the question of the socialist or capitalist genealogy of reform, Deng stated: 'The chief criterion for making that judgement should be whether it promotes the growth of the productive forces in a socialist society, increases the prowess of the socialist state and raises living standards' (Deng 1993, p. 372). Subsequently known as the 'three criteria', Deng's formulation means that any reform is socialist as long as it meets these three criteria. Moreover, Deng sought to distinguish the aims of socialism from the means of achieving it, putting planning and the market on equal footing. He argued:

> A planned economy is not equivalent to socialism, because there is planning under capitalism too; a market economy is not capitalism, because there are markets under socialism too. Planning and market forces are both means of controlling economic activity. *The essence of socialism is liberation and development of the productive forces, elimination of exploitation and polarization, and the ultimate achievement of prosperity for all.* (Deng 1993, p. 373, my emphasis)[4]

Deng's speech paved the way for an official introduction of the concept of 'socialist market economy with Chinese characteristics' (hereafter 'socialist market economy' or SME). The SME was formally ushered in by Jiang Zemin, the then General Secretary of the CCPCC, in his report at the 14th CCP Congress (Jiang 1992). The report states unequivocally that 'the objective of the reform of the economic structure will be to establish a socialist market economy that will further liberate and expand the productive forces'. It further explains that '[B]y establishing such an economic structure we mean to let

---

[4]   English translation here is based on: https://olemiss.edu/courses/pol324/dengxp92.htm (accessed 4 July 2020).

market forces, under the macroeconomic control of the state, serve as the basic means of regulating the allocation of resources, to subject economic activity to the law of value and to make it responsive to the changing relations between supply and demand' (Jiang 1992). The formulation of the SME signals the CCP's commitment to the mobilization and utilization of market forces for the purpose of regulating resource allocation and raising productivity in China.

The 'Three Represents' and the Scientific Outlook on Development

In the pursuit of modernization as the national goal, to demonstrate the superiority of socialism, successive generations of Chinese leaders have sought to find a Chinese way of achieving this goal, rather than merely following the footsteps of developed capitalist countries. This thinking is manifested in the theory of 'Three Represents' proposed by Chairman Jiang in 2001 at a ceremony marking the 80th anniversary of the CCP's foundation. Echoing Deng's 'three criteria', Jiang argued: 'Our party should continue to stand at the forefront of the times and lead the people to advance victoriously. In summary, it must always represent the development requirements of China's advanced productive forces, the direction of China's advanced culture, and the fundamental interests of the overwhelming majority of the Chinese people' (Jiang 2001). While there is a clear continuity between the formulations of 'three criteria' and 'Three Represents', in terms of the focus on productivity and delivering what the people want, Jiang's formulation once again indicates a new importance attached to an 'advanced culture'.

By the mid-2000s, China appeared to have found its own way, manifested in the conception of the 'Scientific Outlook on Development' (SOD). First briefly mentioned in the 'Decision of the Central Committee of the Communist Party of China on Some Issues concerning the Improvement of the Socialist Market Economy' in October 2003, the concept was defined as follows in Chairman Hu's report to the 17th Party Congress in 2007:

> The Scientific Outlook on Development takes development as its essence, putting people first as its core, comprehensive, balanced and sustainable development as its basic requirement, and overall consideration as its fundamental approach. (Hu 2007)

In the spirit of the SOD, Hu's report calls for 'new and higher requirements' for China's development, especially regarding the first Centenary Goal (i.e., building a well-off society in an all-round way). It advocates 'balanced development to ensure sound and rapid economic growth' in the effort to quadruple the GDP per capita between 2000 and 2020. It also includes a call for the promotion of 'a conservation culture by basically forming an energy- and resource-efficient and environment-friendly structure of industries, pattern of growth and mode of consumption'. This reference to a 'conservation culture' effectively introduced the notion of Ecological Civilization or Ecological Progress (*shengtai wenming* in Chinese) into Chinese official parlance for the first time.[5] The introduction of the SOD and Ecological Civilization indicates a broadening of the official conception of modernization and a strong commitment to sustainable development. It also marks a transformation of the official views of the right relationship between man and nature. Previously, this relationship under Chairman Mao was characterized as *rendingshengtian*, meaning that men's will shall prevail over nature. At times, this amounted to a 'war against nature' (Shapiro 2001). In sharp contrast, under the influence of the SOD and Ecological Civilization, a pro-nature mentality has emerged from the mid-2000s. Indeed, it was in this historic context that, as the provincial CCP boss in Zhejiang Province, Xi Jinping wrote in 2005 that 'clear waters and green mountains are invaluable assets' (Xi 2007).

Arguably, the ascent of an ecological rationality resulted from the need to resolve a dilemma that the CCP came to face by the early and middle of the 2000s. The problem was that, while the CCP had since the late 1970s emphasized economic development, this was becoming untenable due to rising environmental problems during the 10th five-year plan (FYP) (2001–05) period and the anticipated inevitable slow-down of economic growth as the economy matured. Pan Yue, deputy director of the State Environmental Protection Administration (SEPA) (2003–07) and later vice-minister of the Ministry of Environmental Protection (MEP) (2008–16), argued in 2004 that 'using a simple GDP indicator over-simplifies the legitimacy of the ruling party. It makes it seem that GDP growth is the only source of legitimacy and is what underpins the stability of our party's rule, which is unscientific' (Pan 2004). He further argued that economic activities could not entirely be controlled by politics so politics should not be held responsible for all economic outcomes. He observed that in the history of the Western countries, numerous economic crises had not caused legitimacy crises for their political orders precisely because such legitimacy is not identified with economic change.

This means that China's leaders came to realize that, to show true superiority, it must also compete on sustainability. This is challenging as China had a late start in environmental protection. It only started to pay attention to environmental problems in the wake of the United Nations Conference on the Human Environment in Stockholm in 1972. Its first National Environmental Protection Conference was convened in August 1973. The conference 'stressed

---

[5]   The original English translation of the report in *China Daily* rendered it as 'conservation culture'. Subsequently scholars have criticized this translation as too narrow and are instead in favour of the term of Ecological Civilization (iCIBA 2008).

the importance of environmental protection to people's well-being and the adoption of a preventive approach to controlling air, water and solid waste pollution' (Zhang and Zhao 2010, cited in Xu and Chung 2014, p. 397). The internalization of this environmental protection rationality resulted in the introduction of the Environmental Protection Law in 1979 and its promulgation in 1989, which led to the establishment of SEPA and subnational Environmental Protection Bureaus throughout the country. SEPA was further upgraded to ministerial status in 1998, despite a concurrent reduction in the number of ministries from 42 to 20 (Jahiel 1997). Under the influence of the SOD, it became the MEP in 2008.

In his influential work on environmental discourses, Dryzek (2013) places China under Administrative Rationalism (AR), a kind of prosaic, reformist discourse that focuses on environmental problem-solving, instead of sustainability. However, researchers from the China Academy of Sciences (CAS), the top national research institution, claim that China has taken off for Ecological Modernization (EM) since 1998, marked by the publication of the 'Ecological Environment Construction Plan' in 1998 and the 'Guideline for Ecological Environmental Protection' in 2000 (Zhang et al. 2007, p. 664). In 2007, CAS published the *China Modernization Report 2007: Study on Ecological Modernization* in a volume of 450 pages (CCMR 2007). According to Zhang et al. (2007), the report is remarkable in two respects. On the one hand, the report provides an unprecedented down-to-earth assessment of China's international standing in terms of EM development. Indeed, using a proprietary Ecological Modernization Index comprising 30 indicators, the report ranks China as 100th out of the 118 countries sampled. On the other hand, the report seeks to refine the concept of EM and to differentiate pathways to EM. It identifies four levels of civilization in an ascending order, including: primitive civilization, agricultural civilization, industrial civilization, and finally, knowledge (ecological) civilization. Furthermore, the report delineates three pathways of EM, namely: (1) 'comprehensive ecological modernization', characterized by dematerialization, decoupling and ecological rationalities; (2) 'integrated ecological modernization' involving accelerated greening of industrialization and the ecologization of the economy towards a knowledge society; and (3) 'ecological modification of classic modernization' involving the modification of conventional industrialization and urbanization according to ecological principles. Zhang et al. (2007, pp. 662–3) suggest that this report 'should be understood ... as an urgent and timely effort to insert ecological rationality into the modernization discourse, policy-making, and practice in China'. It is worth noting that ecological civilization is considered in CAS's analysis as the apex of civilization.

Ecological Civilization

In his report to the 18th CCP Congress in 2012, the departing General Secretary of CCPCC, Hu Jintao, explained the rationality of promoting ecological progress, more commonly known as ecological modernization, as follows:

> Promoting ecological progress is a long-term task of vital importance to the people's wellbeing and China's future. Faced with increasing resource constraints, severe environmental pollution and a deteriorating ecosystem, we must raise our ecological awareness of the need *to respect, accommodate to and protect nature*. We must give high priority to making ecological progress and incorporate it into all aspects and the whole process of advancing economic, political, cultural, and social progress, work hard to *build a beautiful country, and achieve lasting and sustainable development of the Chinese nation*. (Hu 2012, p. 33, my emphasis)

Apart from its intrinsic appeal, three contextual factors may have played an important role in catapulting Ecological Civilization to the top of the Chinese political agenda. These are: (1) its resonance with rediscovered traditional values in the Chinese culture; (2) its supposed Marxist roots; (3) the active role that Chinese scholars have played in further developing this concept and the potential for developing international leadership. The first appearance of the term 'ecological culture' has been traced to an article on *Scientific Communism* published in the Soviet Union in 1984 (Marinelli 2018). When an abridged and translated version of the article appeared in *Guangming Daily*, the word 'culture' was replaced with 'civilization' according to Marinelli. Ye Qianji, a professor based in Chongqing, has been identified as the key figure in the popularization of this term. In his keynote speech at the National Conference on Eco-Agriculture in 1987, Professor Ye highlighted the deterioration of the Chinese ecological environment and called for the vigorous promotion of ecological modernization. Ye once asked '*Rendingshengtian haishi Tianrenheyi*?', which means 'Man's will, no[t] heaven, decides' or 'Man lives in harmony with heaven?' (Marinelli 2018, p. 375). Marinelli observes that the discourse on ecological civilization 'started in the 1980s in the academic domain and was later appropriated by the political discourse' (p. 382). Pan Yue discussed at length the Confucius philosophy of '*Tianrenheyi*' in October 2003 (Pan 2003).

Liu Ren-Sheng has shed further light on the development of this concept. From the Central Compilation and Translation Bureau under the CCPCC and as the lead translator of Roy Morrison's *Ecological Democracy* (Morrison 1995), Liu suggests that this concept carries the genes of both evolutionism and materialism (Liu 2016). Professor Iring Fetscher, a German academic, political scientist and researcher on Hegel and Marxism, is credited by Liu for having coined the term and used it to refer to the civilization succeeding industrial civilization. Among Chinese scholars, Professor Liu Si-Hua is credited for having first proposed the concept of 'socialist ecological civilization' at a national forum on ecological economics in 1986. Two pathways (i.e., strategies) towards Ecological Progress have been proposed by Chinese researchers and policy makers (Liu 2016). These refer to 'the theory of transcendence' and 'the theory of balance'. The former school merely follows Fetscher in considering ecological civilization as a system superseding industrial civilization. According to Liu, its key supporters included: Ye Jianji, Qin Junsheng, Yu Muchang, Yu Keping, as well as Pan Yue (a vice minister at the MEP). The latter school, represented by Professor Liu Si-Hua, agrees with the former on the definition, but stresses the need to balance the relationship between ecological civilization, material civilization and spiritual civilization. Moreover, the latter school promotes a 'five-in-one' strategy, with the five referring to economic development, political development, cultural development, social development and ecological progress. According to Liu (2016), after the 17th Party Congress in 2007, national leaders and key ministerial officials have mainly adopted 'the theory of balance'. Indeed, the 18th CCP Congress report by Chairman Hu (2012) includes the 'five-in-one' formulation as a key strategy for the country and emphasizes that the construction of Ecological Progress should comprehensively run through each and all of the different aspects of the developments of economy, politics, culture, society, and ecology. Finally, Liu Zong Chao, a geologist-turned-ecological activist and the author of a Chinese-language book titled *The Concept of Ecological Civilization and the Trends of China's Sustainable Development* claims that, globally, it is Chinese scholars who have championed the concept of ecological progress and, in doing so, played an irreplaceable role in its development (Zhang 2009). Therefore, Chinese leaders see in the practice of Ecological Progress an opportunity to develop China's leadership in global affairs.

It should be pointed out that the pursuit of Ecological Progress goes beyond mere rhetoric. The choice of 'the theory of balance' is indicative of the policy makers' interest in operationalizing it. In fact, the 'Integrated Plan for Promoting Ecological Progress' (hereafter 'the Integrated Plan') (CCPCC and State Council 2015a) has taken steps to institutionalize the implementation of Ecological Progress. This plan pronounces its aim as building a resource-efficient and environment-friendly economy that is compatible with the goal of building a comprehensively well-off society by 2020, thus integrating the ecological and developmental rationalities. Moreover, it is presented as a strategy for enhancing national prowess, influence and competitive advantage (CCPCC and State Council 2015b). This plan seeks to establish an all-encompassing Ecological Progress supporting system comprising 'source prevention', 'process control', 'damage compensation', and 'accountability'.

The Pursuit of Global Influence and Leadership via Environmental Protection and Climate Actions

Alongside performance legitimacy, productivity- and sustainability-centred discourses that essentially target the domestic audience, the leadership has increasingly felt the need to win external influence in the past decade. This is manifested in the discursive shift from self-determination, indigenous innovations to global leadership. Joshua Ramo (2004) coined the term 'Beijing Consensus' in contradistinction with the 'Washington Consensus' to encapsulate the emergence of this discursive awakening in the early and mid-2000s in China. According to his analysis, the 'Beijing Consensus' contains three theorems about 'how to organize the place of a developing country in the world'. First, it 'insists on the necessity of bleeding-edge innovation' (p. 12). Second, it emphasizes the need to go beyond measures such as GDP per capita and to focus instead on quality of life. It 'demands a development model where sustainability and equality become first considerations, not luxuries' (p. 12). Third, it contains a theory of self-determination. Ramo's characterization is a testament to China's search for a new development model by the early 2000s. The latter is expected to be based on innovations and conducive to self-determination.

In fact, the early 2000s also marked the beginning of a process where well-placed Chinese political actors came to recognize the need and potential for China to innovate on different fronts including environmental protection. As early as 2003, in a keynote speech at the first 'Green China Forum', Pan Yue (Pan 2003), a deputy minister at the SEPA, stated that

> It can be predicted that in the future, the leading countries of ecological industrial civilization will dominate the world structure. Whoever has completed the transformation from traditional industrial civilization to ecological industrial civilization will gain comprehensive advantages in morality, economy, technology, and culture. Those whose environmental problems deteriorate will become passive in international relations. To develop China's environmental culture is not only a response to the environmental crisis we are facing, but also a positive response to the changes in the world and in international relations.

This indicates that, by the early 2000s, some Chinese decision-makers had realized the political significance of protecting the environment and developing an environmental culture. Moreover, coupled with China's continuous economic development in the wake of the global financial crisis that started in 2008 and the assumption of the top leadership position by Xi Jinping in 2012, Chinese elites developed a new kind of global outlook by the early 2010s. Chairman Hu's departing report to the 18th CCP Congress in 2012 proposed a new concept, 'a community of common destiny' (CCD) (*mingyun gongtongti* in Chinese). Hu explains:

> A country should accommodate the legitimate concerns of others when pursuing its own interests; and it should promote common development of all countries when advancing its own development. Countries should establish a new type of global development partnership that is more equitable and balanced, stick together in times of difficulty, both share rights and shoulder obligations, and boost the common interests of mankind. (Hu 2012)

The ascent of this international rationality forms a stark contrast with and break from the earlier Dengist strategy of keeping a low profile (*taoguang yanghui* in Chinese), but is consistent with the pursuit of comprehensive power (Shambaugh 2013). A key moment in this development was Chairman Xi's speech at the National Conference on Propaganda and Ideological Work on 19 August 2013, where he emphasized the importance of ideological work and discursive power (*huayuqua* in Chinese). He called for the consolidation of the guiding position of Marxism in the ideological field and the consolidation of a common ideological basis ('double consolidations' in short) (Xinhuanet 2013). In this speech, Chairman Xi stressed the importance of telling 'Chinese stories', realizing the 'Chinese dream', and sharing 'Chinese solutions'.

In summary, the elite discourses in China have come to encompass numerous rationalities, including a modernist rationality as per the 'Four Modernizations' programme, a socialist rationality as expressed by the FCPs, a techno-economic rationality with emphasis on productivity and economic development, a liberal market rationality in the form of the SME, a sustainable development rationality embodied in the SOD, an ecological rationality through the discourse of Ecological Civilization, and an international rationality expressed by the CCD. As shown above, these diverse rationalities are the result of a series of adaptive steps taken to stay relevant by the CCP. In other words, they are the result of policy learning (Hall 1993). Despite the diversity, however, there is a central logic, namely that fundamentally all policy practices must be consistent with the FCPs and help showcase the superiority of the Chinese socialist system. Deng's message is loud and clear: 'if we want socialism to achieve superiority over capitalism, we should not hesitate to draw on the achievements of all cultures and to learn from other countries, including the developed capitalist countries, all advanced methods of operation and techniques of management that reflect the laws governing modern socialized production' (Deng 1993, p. 373). This logic requires the CCP leadership to continuously improve China's state-led development model and make it compatible with sustainable development, what the Chinese people want and increasingly what people everywhere value. The evolution from 'three criteria', the 'Three Represents', to SOD, Ecological Civilization and CCD is a testament to the CCP's efforts to respond to China's emerging developmental challenges both at home and abroad. Actively promoting MLCT is an integral part of these efforts.

POLITICAL RATIONALITIES FOR MITIGATION AND LOW-CARBON TRANSITIONS IN CHINA

The shifts in political rationalities reviewed above constitute the broad discursive field within which the more specific political rationalities for MLCT have developed in the PRC. In the first ever Comprehensive Work Plan for Energy Conservation and Emissions Reduction (covering the 11th FYP), the State Council (2007b) stated: 'We should take the completion of the task of energy conservation and emissions reduction as an important criterion to judge whether the Scientific Outlook of Development is implemented or not and whether [your] economic development is "good" or not' (p. 2). With an initial focus on mitigation, this work has broadened to embrace systemic low-carbon transition (LCT) from the mid-2010s. Below we will examine the political rationalities converged to promote MLCT by applying Dean's 'analytics of government' (see Table 2.1).

Creating the Field of Visibility

What and who to be governed
The field of visibility for energy conservation and emissions reduction (ECER, or *jieneng jianpai* in Chinese) is built upon what was created for an earlier, related governing objective: energy conservation (EC, or *jieneng*). The latter stemmed from China's national strategy of quadrupling its GDP between 1980 and 2000, as approved by the 12th CCP Congress in 1982 (Zhao and Li 2018). A critical problem for the central planners was that if the energy elasticity remained at 1.58 (a historic average up to that point), quadrupling the GDP would mean that China's total energy consumption (TEC) would have to increase from 0.6 billion tonnes of coal equivalent (tce) in 1980 to 4 billion tce by 2000 – a daunting task; even at the elasticity of one, China's TEC would still be at 2.4 billion tce by 2000. To solve this problem, according to Zhao and Li (2018), researchers at the State Council's Technology and Economic Research Centre studied both theory and practices. By examining relevant data on more than 100 countries, the researchers found that China's energy intensity was the highest in the world, respectively 7 times, 5.8 times and 2.9 times as high as Japan's, France's, and India's. They also concluded that, if energy elasticity could be lowered to 0.5 through implementing 'generalized energy conservation' and a 'comprehensive energy efficiency strategy', it would be possible to realize the target of quadrupling the GDP by only

doubling the TEC. The researchers thus recommended to the government to address energy shortage by improving comprehensive energy use efficiency to lower the economy's energy intensity. This was how China's 'doubling energy use for quadrupling GDP' strategy came about. It led to the formulation of the 'Interim Regulations on Energy Conservation Management' (hereafter 'Interim Regulations') issued by the State Council (1986) (Zhao and Li 2018). The Interim Regulations eventually led to the Energy Conservation Law (ECL), effective from 1998.

The Interim Regulations laid an important foundation for subsequent legislations on EC. They articulated the strategic importance of EC, defined major energy-using units (MEUUs), and introduced an EC responsibility system, an energy statistics system, and a system of setting energy consumption quotas (ECQ) for energy-consuming appliances and equipment (ECAE). The MEUUs were defined by the Interim Regulations as those units consuming at least 10,000 tce a year. However, the ECL expands the definition of MEUUs to include also those units that consume between 5,000 and 10,000 tce a year and are recognized as MEUUs by national and province-level authorities. Moreover, the ECL requires sub-national governments (SNGs) to incorporate EC work into their local socio-economic development plans (SEDPs), the relevant departments of the State Council to formulate EC standards, feasibility studies of fixed asset investment (FAI) projects to take account of rational use of energy, and for county and above governments to compile and publish statistics on energy use. It also includes a compulsory elimination system for energy-inefficient appliances and equipment, and for province and above governments to formulate ECQ for ECAE. Furthermore, the MEUUs are required to submit regular energy use reports and to establish dedicated energy management posts. Finally, individuals are obliged to conserve energy and enjoy the right to report on energy-wasting behaviours. They should be rewarded for contributing to EC and the development and deployment of EC technologies. Those who violate rules on the production, use and transfer of substandard ECAE shall be liable to legal penalties. These provisions make SNGs, MEUUs managers, FAI investors, ECAE producers and users, and individual citizens visible for the purpose of EC. They also spotlight energy statistics, ECQ, ECAE, SEDPs, EC standards, FAI and EC technologies.

A key milestone in developing the *jieneng jianpai* field of visibility was the 11th FYP (2006–10). Under the influence of SME and SOD, this plan (State Council 2006a) featured three historic innovations that have lived on since then. First, it replaced the word 'Plan' (*jihua*) with the term 'Planning' (*guihua*), acknowledging the increased influence of the market economy under SME. Second, it introduced, for the first time, a contradistinction between binding (*yueshuxing*) and anticipative (*yuqixing*) planning indicators and targets. The plan clarified that, while the government at different levels must ensure to meet the binding targets through public resources allocation and other administrative means, it would play only an enabling role to help market players to realize the anticipative targets through market mechanisms. Third and finally, it designated, for the first time, eight binding indicators, none of which was related to the economy. Instead, these binding indicators were exclusively related to society and the environment, including the reductions of the energy consumption per unit of GDP ('energy intensity' in short) by '20 per cent or so' and the reduction of total emissions of sulphur dioxide and chemical oxygen demand (COD) by 10 per cent during the 11th FYP (State Council 2006a). The last two indicators in the 11th FYP gave *jieneng jianpai* its original meaning, with emphasis on EC and the reduction of air and water pollution.

To help fulfil the energy intensity reduction target in the 11th FYP, the State Council (2006b) issued the 'Decision on Strengthening Energy Conservation'. This decision proposed a series of measures for strengthening EC supervision and management, putting emphasis on planning guidance, an EC censorship scheme for FAI projects, additional EC management procedures for MEUUs, and so on. But the most important new measure introduced with long-term impact was an Energy Conservation Target Responsibility System (ECTRS) and an accompanying Energy Conservation Assessment and Evaluation System (ECAES) for SNGs and MEUUs. The various provisions of the decision, including the ECTRS and ECAES, were later incorporated into the ECL 2007 amendment, making SNGs the key agents of managing EC legally. Moreover, the ECTRS and ECAES form a core element in the wider Energy Conservation and Emissions Reduction Target Responsibility System (ECERTRS), designed to help meet the binding MLCT planning targets in the 11th, 12th and 13th FYPs. These systems are explained in detail in Chapter 4.

The ECL 2007 amendment also introduces a number of other governing measures: (1) an industrial policy that restricts the development of high-energy, high-pollution industries (commonly referred to as 'two-highs') and encourages energy-conserving and environment-protecting industries (or 'ECEP industries' in short); (2) the formulation of mandatory energy efficiency standards (MEES) for ECAE and ECQ for energy-intensive production processes by national standardization agencies; (3) the publication of a promotion catalogue for EC technologies and products and a public procurement catalogue of EC appliances and equipment, giving priority to those possessing EC certification. Furthermore, each energy-using unit (EUU) is required to establish an ECTRS. Producers and importers must use energy efficiency labelling (EEL) and register these labels with authorized agencies. MEUUs are now required to submit energy use reports annually, which relevant government departments must scrutinize and, if necessary, instruct audit, and specify time-limited remedial actions in cases of unsatisfactory performance.

MEUUs are required to not only establish energy management posts as under the original ECL, but also appoint energy managers from qualified personnel. Rather than referring to energy use generally, the 2007 amendment specifies rules for the rational use of energy in four separate areas including industry, construction, transport, and public institutions.

The Renewable Energy Law (REL), effective from 1 January 2006, also helps create the field of visibility for encouraging the development of renewable energy (RE). Relaxing the pre-existing restriction for private investment into energy – a strategically important sector, the REL encourages the participation by different forms of ownership in the development of RE, thus opening it up to the private sector. It makes power grid companies liable for fines if they fail to purchase energy generated from RE sources. The REL's 2009 amendment also gives SNGs at or above county-level the responsibility to manage the development and deployment of RE (Article 5). It requires that the relevant administrative authorities for energy under the State Council shall, in accordance with the national energy demand and the actual situation of RE resources, formulate medium- and long-term overall targets for the development and utilization of RE resources, and jointly with the provincial governments, determine medium- and long-term targets for the development and utilization of RE in these sub-national jurisdictions. Because of the ECL 2007 amendment and the REL 2009 amendment, several key subject groups are identified. Moreover, both mitigation and RE are placed on the government's agenda at different levels. Most importantly, the legal responsibility for meeting the planned *jieneng jianpai* and RE targets is firmly placed on the shoulders of SNGs and their top officials.

Problems to be solved
Problematization has evolved. While the key problem in the 1980s and 1990s was energy use inefficiency in the face of insufficient energy supply, the 11th FYP recognized the need to cut down the energy intensity of the economy for the sake of developing a more sustainable economic model, as required by the SOD. Furthermore, along with the introduction of a binding target for reducing CO_2 emissions per unit of GDP ('carbon intensity' in short) in the 12th FYP (2011–15), the problem from the early 2010s has become how to lower both energy and carbon intensities and expand RE capacities while sustaining economic development. By the mid-2010s, the problem of how to sustain subsidies for RE investment and control high curtailment levels of RE capacities emerged.

As mentioned before, the earlier emphasis on EC derived mainly from domestic concerns about energy supply bottlenecks before the end of the twentieth century. Coupled with economic slow-down caused by the 1997 Asian Financial Crisis, the EC effort enabled China to solve its energy supply shortage by the end of the 1990s. It led to the exclusion of an energy efficiency indicator in the 10th FYP. By the mid-2000s, however, the combination of China's accession to WTO in 2001 and the resultant rapid expansion of China's export-oriented manufacturing industry had created explosive growth of TEC and GHG emissions (Zhao and Li 2018). As a matter of fact, 2007 marked not only the publication of the IEA's (2007) first in-depth study of China's energy use and its rapid increase, but also the year when China became the world's largest CO_2 emitter. Moreover, it coincided with the publication of the Fourth Assessment of Climate Change by the IPCC (2007) and the receipt of a joint Nobel Peace Prize by the IPCC in 2007, which put climate change on top of the international agenda.

These domestic and international circumstances created unprecedented pressure for climate mitigation in China by the mid-2000s. From initially viewing it 'as a scientific issue, introduced from abroad and far removed from the concerns of everyday life' and then as a development issue, yet only as part of the environmental protection agenda during the late 1990s to 2006, the government eventually came to recognize it as a national priority for China by 2007 (Stensdal 2012). Pan Jiahua, a senior researcher at the China Academy of Social Sciences and a member of both the National Climate Change Expert Committee and the State Foreign Policy Advisory Committee, has shed a new light on the changes. He characterizes the shifts in the Chinese government's thinking and approach to climate change in five stages, each of which is associated with a distinct set of problem characterization and action strategy. These include disaster prevention (1978–89), involvement in scientific study (1990–7), defending emission rights externally (1998–2006), coordinated development at home and abroad (2007–13), and taking a leadership role internationally (from 2014) (Pan 2018, p. 626). In other words, China initially regarded climate change only as a natural disaster issue (1978–89), then as a reason to conduct scientific research into climate change for the majority of the 1990s. Between 1998 and 2006, it was preoccupied with defending its emission rights internationally. From 2007, however, it has come to recognize the developmental importance of climate change in its own right and strived to balance its climate actions at home and abroad. Pan (2018) observes that China became so committed to its climate action programme by the late 2000s that it did not ask for international financial support for its 2009 pledge to the Copenhagen Accord. The final phase is characterized by the pursuit of international leadership from 2014 according to Pan.

China's National Climate Change Programme (NCCP) (NDRC 2007a), the first among developing countries, affords us a glimpse into the inner thinking of the Chinese elites. The influence of the SOD is evidenced by the fact that the 'Scientific Approach of Development', which is an alternative expression of SOD, is identified as the first of the NCCP's eight guidelines. Summarizing existing research, the NCCP notes that 'The nationwide annual mean air temperature would increase by 1.3~2.1°C in 2020 and 2.3~3.3°C in 2050 as compared with that in 2000' (NDRC 2007a, p. 5). This represents a recognition that warming in China was (and still is) expected to be significantly more serious than the global average. The NCCP identifies a range of problem areas to be tackled, including economic restructuring; energy efficiency improvement; development and utilization of hydropower and other RE resources; ecological restoration and protection; and family planning. Apart from the NCCP (NDRC 2007a), three other documents have also helped clarify the problems. These are the 'National Plan on Climate Change (2014–2020)' (NPCC) (NDRC 2014a), the 'Energy Production and Consumption Revolution Strategy' (EPCRS) (2016–30) (NDRC and NEA 2016a), and the white paper on 'China's Energy Development in the New Era' (CEDNE) (State Council 2020). While the NCCP echoes the SOD, the NPCC and EPCRS reflect the effects of China's Cancun Pledge and the nationally determined contributions (NDCs) under the Paris Agreement, respectively. Finally, CEDNE incorporates the 2060 carbon neutrality goal.

It is evident that under the influence of the notion of Ecological Civilization and differently from the reactive tone of the NCCP, the NPCC has taken the task of addressing climate change to a higher and more strategic level. It sets out three key strategic requirements regarding climate-related work: (1) to regard 'actively addressing climate change' as a 'major national strategy'; (2) to regard 'actively addressing climate change' as 'a major measure of constructing ecological civilization'; and (3) to give full play to the leading role of climate action in related areas including EC, development of non-fossil fuels, ecological construction, environmental protection, and disaster prevention and reduction (NDRC 2007a, p. 3). In other words, from the NPCC in 2014, the work of addressing climate change has evidently become a strategic lever for achieving the Ecological Civilization and promoting development in numerous related areas in China. This represents a step up from mere climate mitigation to systemic LCT.

Objectives to be met
As consistent with the wider political rationalities, the principal ends of carbon government in China are twofold: meeting its international mitigation pledges while continuing to push forward its modernization and Ecological Civilization programmes. A key characteristic of China's MLCT experience is that it has aligned its policy objectives with its international pledges and integrated them into general planning frameworks, especially the FYPs. Fortuitously, the timeframes of the Cancun Pledge and the Paris goals coincide with the two 'centenary goals' of China's modernization programme. This may have given China's MLCT programme an added impetus. Interestingly, by 2020, Chinese

leading researchers had accepted the need to fulfil both the modernization goal and the Paris goals (He 2020). On the other hand, China has increased the number of areas where explicit objectives for the purpose of MLCT are set.

The NCCP (NDRC 2007a) includes four key objectives: (1) to reduce energy intensity by about 20 per cent by 2010; (2) to raise the proportion of renewable energy in primary energy supply to 10 per cent by 2010; (3) to improve resource utilization efficiency and strengthen emission control of nitrous oxide so as to maintain it at the 2005 level; and (4) to increase forest coverage rate to 20 per cent and carbon sink by 50 million tonnes by 2010 over the levels of 2005. It is worth noting that this set of objectives does not include an objective on carbon intensity reduction. Two years later, China's Cancun Pledge (see Table 1.1) included the objective of reducing carbon intensity by 40–45 per cent from 2005 to 2020. This was followed by the inclusion of a carbon intensity reduction target in the 12th FYP (2011–15) (see Chapter 4).

The NPCC (NDRC 2014a) further builds on these objectives. On top of China's Cancun Pledge, it adds three other specific targets for 2020 including: (1) to limit the primary energy consumption (PEC) to 4.8 billion tce; (2) to raise the share of strategic emerging industries (SEIs) to 15 per cent of GDP; and (3) to increase the share of service industries in GDP to above 52 per cent. Significantly, the first target here introduces an absolute limit on PEC in the country for the first time. Compared with previous commitments expressed in terms of the reduction of either energy or carbon intensity, this absolute cap indicates a hardened determination in China's mitigation plans. On the other hand, the second and third targets demonstrate again that the field of visibility has expanded from EC to a wider transformation of the economy, that is, LCT.

To ensure the fulfilment of China's NDC (see Table 1.1) and longer-term sustainable development, the EPCRS outlines policy targets for three stages. First, by 2020, the targets are to limit total energy use to 5 billion tce, to raise the share of non-fossil fuels (NFFs) to 15 per cent, to reduce carbon intensity by 18 per cent and energy intensity by 15 per cent from 2015 levels, and to increase the energy self-sufficiency level to above 80 per cent. Second, during 2021–30, the objectives are to limit total energy use to 6 billion tce, to raise the share of NFFs to around 20 per cent and the ratio of natural gas to around 15 per cent, to reduce carbon intensity by 60–65 per cent from the 2005 level, to reach peak CO_2 emissions around 2030 (and strive to peak early), and to bring energy intensity in line with 'current world average' and energy efficiency in line with advanced international levels in key industrial products. Third and finally, looking forward to 2050, the targets are to achieve a stable amount of energy consumption and to increase the share of NFFs to above 50 per cent. The EPCRS also introduces two new tools for controlling TEC: (1) to implement 'double control', which means controlling both TEC and energy intensity in relevant plans; (2) to establish an energy use allowance system (see Chapter

4 for details). As discussed in Chapter 1 (see Table 1.1), enhanced objectives were announced in December 2020.

It is worth noting that, through the pursuit of MLCT, the policy makers have sought to create new engines of economic growth and to find an effective way of upgrading the economy. By the mid-2010s, economic slow-down had become 'the new normal'. First raised by President Xi in a speech at the 2014 APEC CEO Summit in Beijing, this notion refers to three key characteristics of the current patterns of economic development in China – 'a slowdown in the rate of growth, optimisation of economic structure, and shift of growth engines' (Xi 2017b, p. 251). Reflecting such a reasoning, the 13th FYP (2016–20) puts emphasis on the notion of innovative, coordinated, green, open, and inclusive development. President Xi (2017b) pointed out that 'Green, circular and low carbon-development … guides the direction of the current revolution in science, technology and industry' and could give rise to 'many new engines of growth' (2017b, p. 218). Therefore, the top leadership has regarded MLCT as integral to its ecological rationality as well as modernization and economic development rationalities.

The MLCT programme has also become aligned with China's new 'going global' agenda. The announcement of China's NDCs illustrates this. Significantly, these first announced in a joint statement between President Xi and President Obama of the USA on 25 September 2014. According to Pan (2018), this marked the beginning of a new era for China: addressing climate change has become a strategy of seeking international leadership from that point. In Chairman Xi's (2017a) report at the 19th Party Congress, he made it clear that China's modernization means modernization of the harmonious coexistence of nature and human beings. He declared that it was 'incumbent on China … to take the lead in international cooperation on climate change responses and to engage in, contribute to and lead the efforts to build a global ecological civilisation' (cited in Pan 2018, p. 530).

Forms of Knowledge Utilized and Produced

Knowledge-power, that is, the coupling between the exercise of power and the creation and use of knowledge, is a crucial element in China's carbon governmentality. It is combined with experimental piloting. This covers various branches of social and natural sciences including (but not limited to) climate science, energy use statistics, GHG inventories, energy transition scenarios, LCT pathway analysis, and political economy about MLCT at home and abroad.

Utilizing sciences

The Chinese authorities have invested heavily in information gathering and knowledge creation about its governing objects and subjects, ranging from climate science, energy use, and emission patterns to green credits and RE curtailment rates. This is because compiling energy use statistics and GHG inventories at both national and sub-national level is essential for creating the field of visibility, operationalizing both the ECTRS and ECERTRS at the local level, preparing the ground for energy use allowance (EUA) and carbon emission trading, and engaging with international partners. The compilation of energy use statistics was mandated from the 'Interim Regulations' in 1986. It has led to the development of a nationwide online energy reporting system, with 293 prefecture-level and above municipalities serving as the lowest organizing administrative units (Xu et al. 2016). Preliminary GHG inventory compilation work was first undertaken nationally in 2007 as part of the effort to develop sub-national climate action plans, in correspondence with the NCCP (NDRC 2007a). As will be explained in Chapter 4, a series of efforts have been made to generate local GHG inventories across key sectors since the early 2010s. China published three National Climate Assessment Report in 2006, 2011 and 2015 and initiated the preparation of its fourth one in January 2018.

Overall, China has paid much attention to statistics, which Foucault called 'the science of the state'. Apart from energy statistics, it introduced 'climate change statistics' to the national statistics system from the mid-2010s (NBS 2015). This consists of climate change statistical indicators and activity indicators in the five sectors identified by the NDRC (2011a), including energy activities, industrial production processes, agriculture, land-use change and forestry, and waste treatment. The climate change statistics cover 36 individual indicators under 5 categories and 19 sub-categories. The relevant departments under the State Council are required to submit annual synthesis reports and five-yearly survey reports on climate-related activities (NDRC 2016, pp. 102–3). Technical knowledge has also been developed in other areas including changing technical standards and sectorial benchmarks for EC; RE development and curtailment rates across provinces (see Chapter 4); green credit performance by banks (Chapter 5); cost-benefit analysis of different mitigation options and family carbon footprints (see Chapter 6).

There have been persistent and coordinated efforts to analyse potential energy transition and LCT pathways. For example, researchers from Renmin University and the Energy Institute, NDRC, conducted coal consumption scenario analysis for all 293 prefecture-level or above municipalities from 2010 to 2015 (Xu et al. 2016). The effects of controlling total coal consumption and appropriate policies for cities under 13 categories were examined. One of the latest research projects explored 'long-term low-carbon development strategy and low-carbon transition pathways to 2050' (hereafter LTLCDSLCTP). Run from the end of 2018 to June 2020, this study was led by Tsinghua University's Institute of Climate Change and Sustainable Development and explored 18 different topics. It involved eight top-level research institutions including the State Information Centre, Energy Institute under NDRC, CAS, China Academy of Social Sciences, NCSC, plus three research institutes belonging to the Ministry of Commerce, Ministry of Transport, MEE and four other research institutes from Tsinghua (He 2020). The study suggests that China can strengthen its NDC for 2030. The recommendations include reducing carbon intensity by 'more than 65 per cent', setting the target for NFF's ratio at 'around 25 per cent', striving to reach peak CO_2 'before 2030' and increasing forest stock by 5.5–6.0 billion m^3. A comparison with Table 1.1 shows a high level of similarity between this set of recommendations and what President Xi announced at the UN Climate Summit in December 2020.

Political economy about MLCT at home and abroad
Consistent with performance legitimacy, Chinese state actors have exhibited a strong interest in assessing the system and the officials' performance, often by engaging in international comparison. Back in 1992, Deng noted that 'The economies of some of our neighbouring countries and regions are growing faster than ours. If our economy stagnates or develops only slowly, the people will make comparisons and ask why' (Deng 1993, p. 375). Similar mentality has motivated an increasing emphasis on performance assessment and learning in MLCT. As mentioned earlier in this chapter, China came to develop its 'doubling for quadrupling' EC strategy after learning about its laggard performance in energy use efficiency (Zhao and Li 2018). Explaining the importance of developing an environmental culture, Pan Yue (Pan 2003), the deputy director of the SEPA, berated China's performance in environmental protection, acknowledging that one-third of the country's land had become acidified; air in three-quarters of the cities monitored was unclean; half of the most polluted cities in the world belonged to China; and air pollution was causing 15 million people to suffer from bronchitis (citing UNEP here). For another example, China's carbon emissions trading (CET) pilots build upon its earlier emissions trading pilots on SO_2, which in turn were based on learning from the operation of similar systems in the USA (Zhang et al. 2017).

Moreover, there have been sustained efforts to improve the practice of measuring and evaluating cadre performance on environmental (and social) matters. The cadre evaluation system (CES) and target responsibility system (TRS) have played an important role in ensuring local compliance with central will since their introduction in 1988 (Whiting 2004). While the CES covers five dimensions – virtue, competence, diligence, achievements, and absence of venality, the TRS relates specifically to the dimension of achievements, 'where weight is assigned to performance targets, and local leaders are rewarded or punished based on the fulfilment of the targets set down in the performance contracts agreed with their superiors' (Zuo 2015, p. 960). An outstanding example of such effort was the 'green GDP accounting' project, jointly undertaken by the SEPA and several other bodies including the National Bureau of Statistics (NBS) between 2004 and 2006. It was designed to improve the CES and represented one of the most ambitious experiments in the world in measuring green GDP (Pan 2004). In the end, the project was scuttled due to SNGs' resistance as well as technical difficulties (Li and Lang 2010). Nevertheless, as a result of the project, 'the general idea of using bureaucratic incentives to motivate a greater focus on environmental objectives had taken hold' (Wang 2013, p. 392).

Zuo's (2015) study shows that categories of performance targets were modified across national guidelines issued in 1988, 2006 and 2009. In particular, 'resource consumption' and 'resource conservation' were introduced in 2006, in accordance with the SOD. On the other hand, the regulations announced in July 2009 (CCPCCOD 2009) mandate the assessment of local leading corps at county and above levels in meeting annual targets in three areas: economic development, social development, and sustainable development. The last area covers energy conservation, emissions reduction and environmental protection. The 2009 change obviously reflects the inclusion of the *jieneng jianpai* binding planning targets from the 11th FYP.

The Integrated Reform Plan for Promoting Ecological Progress (CCPCC and State Council 2015a) is a culmination of these reflective efforts. It sets out to establish eight supporting systems: (1) a property rights system of natural resources assets; (2) a land and space development and protection system; (3) a spatial planning system; (4) a total resource management and overall conservation system; (5) a paid use of resources and ecological compensation system; (6) an environmental governance system; (7) an environmental governance and ecological protection market system; and (8) an Ecological Progress performance evaluation and accountability system. This plan attaches great importance to environmental protection and MLCT. For example, under the fourth point, it stresses the need to establish a system of managing the TEC and EC, the gradual cancellation of subsidies for fossil fuels, and the establishment of a national system controlling and disaggregating carbon emissions. Under the eighth point, it calls for the development of a green development indicators system, the development of a warning mechanism for resources and environment carrying capacities, exploratory compilation of natural resource balance sheets, the introduction of an end-of-term natural resource assets audit for cadres, and the establishment of a system of lifelong accountability for causing ecological and environmental damage.

There is much interest in what others are doing abroad. This reflects a genuine desire for learning. China's REL was drafted with the help of Energy

Foundation China (2019), an American charity organization. Knowledge diffusion from abroad and domestic innovation has played an important role in propelling the explosive growth of China's green bonds market since 2016 (Elliot and Zhang 2019). For another example, the LTLCDS, mentioned earlier, benchmarks the Europe Green Deal. It notes that the EU's emission strategy is foremost a 'new growth strategy', aimed at 'realising social equity and prosperity, economic modernization, net zero carbon emissions, resource efficiency and competitiveness in EU by 2050' and 'transiting economic growth towards a more sustainable direction to enable EU to occupy a leading position in the world' (He 2020). It also compares the investment requirements of China and the EU for 1.5°C compatible energy pathways between 2020 and 2050.

Experimental piloting

China's approach to MLCT is characterized by a widespread enthusiasm for experimentation or piloting (*shidian* in Chinese). Again this builds on past experience. Heilmann (2008) identifies a distinctive policy cycle in China, 'experimentation under a hierarchy', and considers it as key to understanding the Chinese system of adaptive authoritarianism. He attributes the system's adaptiveness to open-ended and innovative policy experimentation, in which 'experimenting units try out a variety of methods and processes to find imaginative solutions to predefined tasks or to new challenges that emerge during experimental activity' (Heilmann 2008, p. 3). A key feature of this process is that innovating through implementation comes first and codification of universal laws and regulations comes later. The relationship between the 'Interim Regulations' issued by the State Council (1986) and the ECL (1997) fits this pattern perfectly.

The Low-Carbon City Pilots (LCCP) programme offers a good illustration of this aspect. The learning is first of all reflected in how the pilots were chosen. According to the NDRC, the first batch of eight pilots were chosen mainly based on the localities' enthusiasm. In comparison, the second batch were selected based on expert screening. Moreover, the 45 pilots of the third batch, chosen out of a total of 52 applications, had to demonstrate advanced targets at least in some areas and systemic innovation (21st CBH 2016). Second, by setting application parameters and conducting annual assessment and evaluation, the NDRC has attempted to nudge the cities to learn and to share this learning more widely. To illustrate, the requirements for the first-batch pilots were quite general. The pilots were simply tasked to: (1) prepare low-carbon development plans; (2) formulate supporting policies for low-carbon and green policies; (3) speed up the development of low-emission industrial systems; (4) establish a GHG statistics and management system; and (5) actively promote a low-carbon and green lifestyle and consumption mode (NDRC 2010). However, by 2012, the 32 pilot cities surveyed by Wang et al. (2015) were working towards four key indicators: CO_2 intensity reduction, proportion of NFFs in PEC, forest coverage ratio, and carbon discharge peak time. Furthermore, the third-batch pilots were required to benchmark themselves across 14 indicators (21st CBH 2016). They also had to identify from one to five specific areas for systemic innovation. Third and finally, learning is followed through. So specific milestones were set for the third batch from February 2017 to 2020 (NDRC 2017b). A book published by the Green Low-Carbon Development Think Tank Partnership (GLCDTTP) (2016) features *jieneng jianpai* experiences from 14 LCCP cities.

Perhaps most importantly, through various pilot schemes, China has accumulated a good deal of knowledge about how market mechanisms like the emissions trading system (ETS) work and how LCT can be combined with local economic development. A crucial aspect of this endeavour is experimental activities led by SNGs (see Chapters 5 and 6). Chapter 4 will review, among other matters, a selection of piloting schemes.

Identities and Agency

While disciplinary power is focused on direct command and control on behaviours, governmental power aims to exploit the subjects' own capacities to achieve the transformation of their conduct as desired by the governing authority. Foucault (1982) identifies three modes by which human beings are made subjects: (1) when someone becomes the object of scientific inquiry; (2) being divided inside or from others; and (3) aligning their behaviours with the objectives of the governing authority. These suggest different ways of constituting subjects, ranging from categorization and statistics to awareness campaigns. Moreover, economic incentives, social identification and lifestyle choices can be important levers. China's carbon governmentality has identified a range of key subject groups and sought to provide them with certain statuses, capacities and incentives so that they become responsible and effective carbon managers.

The constitution of key subject groups
As shown earlier in this chapter, through laws, regulations and FYPs, the Chinese state has identified a range of subjects for carbon management. These include (by order of priority): SNGs, EUUs (especially MEUUs), producers and suppliers of ECAE, RE investors, power grid companies, and individuals. Households and financial institutions (FIs) have also been targeted. However, we will deal with them in Chapters 4 and 5. Of all the non-financial groups, the SNGs play a pivotal role. So the discussion in this part will focus on them.

Let us first clarify the administrative system on mainland China. It comprises five levels from the top to the bottom: the centre represented by the

State Council, 31 provinces (including autonomous regions and four centrally administered municipalities – Beijing, Shanghai, Tianjin and Chongqing), 334 prefectures (including 294 prefecture-level municipalities), 2,851 county-level units (including 964 urban districts and 363 county-level cities), and 39,888 townships (NBS 2019). In this book, I use SNGs to refer to all kinds of government below the central level, unless otherwise specified. The prefecture-level or above municipalities represent the most critical tier in the MLCT governing system, because they, totalling 293, accounted for 97.5 per cent of the country's GDP and 95.5 per cent of its energy consumption as of 2010 (Xue et al. 2016). This is also because the Chinese administrative system is generally characterized by the leadership of municipalities over counties (*shi dai xian* in Chinese). Municipalities are the lowest administrative units in energy statistics. Moreover, cities represent the mainstay of the various pilot schemes (see Chapter 4). The SNGs have dual identities. They act as governors relative to other local subject groups, but also as the governed from the viewpoint of the central government. This means that, on the one hand, they are subject to central control under the ECTRS, ECERTRS and the like. On the other hand, they have their own interests and agency thanks to their stake in the political economy. The latter is explained below.

Status, capacities and incentives
Much emphasis is placed on the legal responsibilities of various subject groups. In particular, in comparison with the original ECL, the 2007 ECL amendment increases the number of clauses under the chapter on 'legal responsibilities' from 8 to 18. These spell out penalties for failing to abide by the rules related to the censorship for FAI projects, the elimination system (for substandard productive capacities), EC certification labelling, the ECQ, power grid companies' obligation to connect RE sources, misconduct in public procurement and the management of MEUUs. The 2016 ECL amendment further tightens the FAI projects' EC audit and censorship. On the other hand, the 2007 amendment of the ECL introduces an entire chapter on incentive measures compared with the original 1997 ECL. These measures include the establishment of special funds for EC within central and provincial budgets, taxation incentives for relevant technologies and products, subsidies for energy-conserving products, favoured public procurement treatment of EC appliances and equipment, credit support and differential energy pricing by time and sector, and a system of recognition and rewards. Under the ECL, individuals as citizens and consumers are endowed with five different roles. Apart from their obligation to conserve energy, rights to report energy-wasting behaviours, and entitlement to rewards for contributing to EC, they should be guided towards the use of energy-saving technologies and products and be incentivized to save energy through pricing

mechanisms. As discussed earlier, the EUUs (especially MEUUs) bear similar, but more stringent obligations and responsibilities.

In particular, the 2007 ECL amendment has substantively increased the legal responsibilities of SNGs. They are required to incorporate the work of EC not only into local SEDPs, but also into annual plans. They must formulate and implement medium- and long-term EC plans and annual plans. Moreover, SNGs are required to report EC work annually to local legislative bodies. Most critically, the 2007 amendment puts the ECTRS and the ECAES on a legal footing which requires that the outcomes be incorporated into the existing annual CES. The 'Integrated Plan' (CCPCC and State Council 2015a) has further entrenched the administrative role of SNGs. Section 9 calls for the establishment of an Ecological Civilization Responsibility System for leading cadres during their terms and the perfection of the ECERTRS. Moreover, for the first time, it introduces a system of lifelong accountability. This means that a cadre will be held responsible for causing serious damage to resources, ecology and the environment either wilfully or out of neglect, even after the end of their office term.

These requirements are changing the central–local relationship in China. The Chinese governance system has been characterized as a 'regionally decentralized authoritarianism' (RDA), combining highly centralized political and personnel controls at the national level and a regionally decentralized administrative and economic system (Xu 2011). In the earlier growth-oriented regime, RDA was credited with solving the principal-agent problem inherent in a hierarchical political system by encouraging officials to engage in tournament-like inter-region competition on GDP growth; they also competed on testing new ideas and policy practices. Some researchers have attributed China's developmental state and China's extraordinary post-reform economic success to RDA (e.g., Xu 2011; Knight 2014). Regardless of the debate on the merits of RDA, it has given SNGs a key role in governing economic activities in their jurisdictions. Today, they still wield significant power as full or partial owners of a large number of state-owned enterprises (SOEs) (including financial institutions).

However, the fiscal reform in 1994 drastically recentralized China's public finance and has made SNGs heavily dependent on central transfers or borrowing financially. The Budget Law issued in August 2014 and effective from January 2015 has further increased central control over local public finance. Now SNGs can only borrow by issuing bonds, the amount of which is subject to annual provincial quotas allocated by the State Council. The Budget Law also sanctions that the comprehensive budget must encompass all public budgets including: General Public Budget (GPB); Government Fund Budget (GFB) (including revenues from land sales); State Capital Budget (SCB) (related to the operation of SOEs); and Social Security Fund Budget. An analysis of the 2018 final national public budget shows that SNGs accounted for 53.4 per cent (RMB 9,790.338 billion) of total GPB revenue, but 85.2 per cent (18,819.632 billion) of total GPB expenditure, suggesting heavy reliance by SNGs on central transfers for filling the gap. However, SNGs' shares in total GFB revenue and SCB revenue were 94.7 per cent and 54.3 per cent respectively, representing RMB 7,137.2 billion and RMB 157.5 billion, respectively. Moreover, earning from transfers of land-use rights by SNGs was as high as RMB 6,509.6 billion, or 91.2 per cent of their GFB revenue, or an equivalent of 66.5 per cent of local GPB revenue. These figures testify to the crucial importance of land-based finance to Chinese SNGs (see Xu 2019 for more details).

In this context, SNGs face two principal MLCT strategies: one based on LCT-driven industrialization and the other on new forms of urbanization. Kung et al. (2013) suggest that because of the fiscal reform since 1994, the incentives for SNGs have shifted away from industrialization to urbanization, as SNGs have an exclusive claim on real estate-related taxation and fees, but must share a majority of other corporate taxation revenue with the centre. Under this condition, successful low-carbon industrial development can reduce the cost of mitigation for the locality, which is what has happened in Rizhao (Huang et al. 2018), and bolster the local economy, as in the case of BYD Auto for Shenzhen (Xu and Chung 2014; Huang et al. 2018; Huang and Castán Broto 2018). However, MLCT will also inevitably incur costs through the phasing out of substandard productive capacities, slower economic growth and associated job losses. Meanwhile, urbanization offers both opportunities and challenges for LCT. While urban growth makes it easier to create green buildings and introduce lower-carbon infrastructure, more compact development and lower carbon footprints under MLCT could spell the end of a land-intensive development model, which has for a long time underpinned municipal finance in many cities (Xu and Chung 2014; Xu 2016). A combination of slower economic growth and restrained spatial growth could reduce both budgetary revenues and non-budgetary revenues, especially fees from land-use rights transactions, further constraining local financial resources.

In summary, while the discourses around the SOD, Ecological Civilization, 'the new normal' and the exclusion of economic indicators from the binding planning targets have provided some incentives towards MLCT for SNGs, economic and financial rationalities could lead them onto other paths. Chapter 6 will examine how three Chinese SNGs have responded to the centrally crafted carbon governmentality and developed their own.

CONCLUSION

Thanks to its Marxist tradition, discourses act as a potent strategic force in the PRC. While the hegemonic socialist discourse still dominates, the governing elites have evidently acquired a multiplicity of rationalities, ranging from the modernist and techno-economic, to sustainable, environmental, and ecological rationalities. They are also influenced by both market and planning rationalities. Overall, motivated by a socialist rationality and the desire to maintain political legitimacy, the elites have enriched the socialist and modernistic rationalities by incorporating an ecological rationality since the mid-2000s. Moreover, the past decade has witnessed the institutionalization of Ecological Civilization from 2012 and the onset of internationalism under the leadership of President Xi from 2014.

The chapter has shown that combined, these wider rationalities have had the effects of supporting *jieneng jianpai* and LCT in China. In other words, the MLCT programmes have benefited from the transformation of the wider Chinese governmentality from merely socialist and techno-economic to a hybrid governmentality that combines socialist, neoliberal, biopolitical, ecological and international rationalities. A key finding is that the MLCT rationality resonates to a large extent with the pre-existing discourses on accelerated accumulation, reform and opening-up, the SME, and the emerging discourses on the SOD, Ecological Civilization and CCD.

This chapter sheds light on the field of visibility, the forms of knowledge utilized and produced, and the forms of identities and agency prominent in the MLCT programmes. First, carbon governmentality in China has rendered a range of actors and emission sources visible and governable. It is targeted at eight key groups – SNGs, MEUUs, manufacturers and suppliers of ECAE, power grid companies, RE developers, FIs, households, and individuals. The role of SNGs is especially central in this system thanks to their pivotal position defined in the ECL and REL and the institutional capacities that they have developed in the past decades. They also play an important role in the creation of local knowledge base about GHG sources and MLCT options through their experimentations and participation in low-carbon pilots. Chapter 6 will shed further light on these.

Finally, it is evident that the centre is trying to turn these targeted groups into responsible and effective carbon managers through a range of means, including laws, regulations, standards, sanctions, information, knowledge, and economic incentives. It is discernible that there is potential tension between the centre and SNGs, because of their differing standpoints and interests. While the centre and localities share an outlook towards the maintenance of existing socio-economic and political systems and the techno-economic rationality, they differ in important aspects. Specifically, while SNGs are concerned with local economic development, land-based finance and the preservation of local ecological resources, the centre appears to be more concerned with social stability and inter-system, inter-state competition.

The next chapter will explore what kinds of governing technologies have been utilized for the purpose of promoting MLCT in China.

4. Beneath China's low-carbon transitions: governing techniques and technologies

INTRODUCTION

As one of the oldest centralized states, China is no stranger to the concept of the art of government. After all, China created the world's first civil service. To achieve mitigation and low-carbon transitions (MLCT), however, the government must go through a real struggle to change not only the minds, but also the behaviours of some powerful groups, who have for long focused on an economic development agenda. The purpose of this chapter is to analyse the set of technical means that the state has mobilized to do so. Building on Chapter 2, this chapter makes a distinction between governing techniques and technologies (GTTs). Compared with techniques of government, technologies of government are distinguished by their structural complexity, inter-system connectivity, force relations and an orientation towards performance optimization. Following the discussion in Chapter 3, we will concentrate on the GTTs that have been applied to the seven non-financial key subject groups. Furthermore, the focus will be on energy conservation (EC), although renewable energy (RE), greenhouse gas (GHG) mitigation, and climate-friendly equipment manufacturing activities are also covered. This is because more than 80 per cent of carbon emissions reduction in China is attributed to EC (Yu and Meng 2017). Xie Zenghua, China's special envoy on climate change, claimed in 2018 that China's cumulative energy saving since 2005 had accounted for more than 50 per cent of the world's total (SCPO 2018). This chapter will also examine the GTTs' effectiveness on MLCT outcomes when feasible.

To guide the conduct of the targeted subject groups, the central government has assembled a wide range of GTTs, including most importantly the following: (1) comprehensive planning; (2) target responsibility systems; (3) re-organization of the bureaucracy; (4) re-ordering of performance criteria and strengthened accountability for government officials; (5) compilation of GHG inventories; (6) experimentation and piloting; (7) mandatory energy efficiency standards (MEES) and RE requirements; (8) market-based mechanisms; (9)

financial incentives and investment; and (10) training, education and capacity building. Below we will examine these.

COMPREHENSIVE PLANNING AND TARGETS DISAGGREGATION

China's most distinctive technique of government is no doubt planning. Thanks to its socialist legacy, China has a comprehensive system of planning, characterized by a '4 by 3' structure. On the one hand, there are four kinds of planning: overall planning, spatial planning, special (sectoral) planning and regional planning. On the other hand, statutory planning is conducted at three administrative levels: nation, provinces, and municipalities and counties. In addition, a balance of short-, medium- and long-term planning are conducted. The linchpin of this system is the social and economic development five-year plans (FYPs) (Zhu 2018).

FYPs

The FYPs have played a central role in promoting MLCT. However, both their scope and strength have evolved. This refers to the fact that while successive FYPs, except the 10th, have set energy efficiency targets, albeit for different reasons, more recent FYPs have added new targets and made these targets more stringent. As discussed in Chapter 3, the 11th FYP (2006–10) was a key turning point, prompted by the newly introduced Scientific Outlook of Development (SOD). It marked the downgrading of economic development objectives by setting the eight binding planning targets exclusively related to environmental and societal matters. Five of these binding indicators were environment-related: (1) reducing energy consumption per unit of GDP ('energy intensity' in short) by 20 per cent; (2) reducing industrial water use per unit of GDP by 30 per cent; (3) limiting the reduction in protected arable land to 30 million hectares; (4) reducing total emissions of sulphur dioxide and chemical oxygen demand (COD) by 10 per cent; and (5) increasing forestry coverage by 1.8 per cent (State Council 2006b). Furthermore, following the 2009 Copenhagen Pledge (i.e., the Cancun Pledge) (Su 2010), apart from the energy intensity reduction target, the 12th FYP (2011–15) added two new MLCT-related binding targets: (1) raising the share of non-fossil fuels (NFFs) in primary energy consumption (PEC) to 11.4 per cent; (2) reducing CO_2 emissions per unit of GDP ('carbon intensity' in short) by 17 per cent.

Furthermore, the 13th FYP (2016–20) included the same set of binding indicators as those in the 12th FYP (see Table 4.1). In addition, the 13th FYP for Energy Development (2016–20) (NDRC and NEA 2016b) went one step further, adding two new binding indicators for 2020. The first was to limit the proportion of coal in total energy consumption (TEC) to 58 per cent (i.e., a reduction of six percentage points), while the second was to reduce the level of coal consumption by coal-powered generators from 318 gsc/kWh to less than 310 gsc/kWh. It also included two new anticipatory energy-related targets: (1) limiting TEC to less than 5 bn tce; (2) limiting the total coal consumption to 4 bn tonnes of crude coal. The imposition of a TEC cap, though not strictly binding, alongside the energy intensity control target is called 'double-control'. The latter signals an important transition from a sole reliance on intensity control to both intensity and quantity control, thus paving the way towards controlling total carbon emissions. Finally, the 14th FYP announced in March 2021 sets the energy and carbon intensity reduction targets at 13.5 per cent and 18 per cent respectively (see Table 4.1). While the energy intensity target is lower than those in the previous three FYPs, possibly because of diminished scope for further efficiency improvement, the carbon intensity reduction target is the same as that in the 13th FYP. Notably, the 14th FYP outline no longer includes a binding target for the share of NFFs in total TEC. However, President Xi's announcement committed China to a share of around 25 per cent by 2030 (see Table 1.1). As shown in Table 4.1, by gradually incorporating and strengthening energy and carbon intensity reduction and RE planning targets, China's last three FYPs have put the MLCT firmly at the top of the government's agenda.

Also from the 11th FYPs, the Chinese government has formulated a special work plan for *jieneng jianpai* to ensure that the binding planning targets are met. The State Council launched a series of Comprehensive Work Plans (CWPs) for *jieneng jianpai* for the last three FYPs (State Council 2007b, 2011a, 2016b) and a Work Plan for Controlling Greenhouse Gas Emissions (WPCGHGE) for the last two FYPs (State Council 2011c, 2016d). The CWP is where the targets are disaggregated for implementation. Finally, the National Development and Reform Commission (NDRC) monitors annual progress towards these targets, and issues additional work programmes when necessary, as it did in 2007 and 2014. For example, in 2014, the State Council introduced an additional action plan when it realized that certain planning targets were not on track (State Council 2014). This action plan outlines 30 counter-measures and specifies, province-by-province, detailed targets for eliminating old coal-fired boilers, reducing air pollutants and phasing out old vehicles. Annual monitoring of RE development and utilization has been initiated by the National Energy Administration (NEA) since 2017.

Table 4.1   Climate-related binding planning targets in the 11th, 12th, 13th and 14th FYPs

FYP (period)	Reduction of energy use per unit of GDP (%)		Share of non-fossil fuel in primary energy consumption (%)		Reduction of CO_2 emissions per unit of GDP (%)		Share of coal in energy consumption (%)		Increase in forest coverage (%)		Increase in forestry stock (bn cubic metre)	
	Target	Actual	Target	Actual	Target	Actual	Target	Actual	Target	Actual	Target	Actual
11th (2006–10)	20	19.1	n/a	n/a	n/a	n/a	n/a	n/a	1.8	2.16	n/a	n/a
12th (2011–15)	16	18.2	11.4	12	17	20	n/a	n/a	1.3	n.d.	0.6	n.d.
13th (2016–20)	15	14.3**	15	15.9*	18	19.5**	58	57.7 (end of 2019)	1.38	n.d.	1.4	n.d.
14th (2021–25)	13.5	n.d.	20	n.d.	18	n.d.	n.d.	n.d.	0.9 (from 2019)	n.d.	n/a	n/a

Note: *As percentage of total energy consumption. **Expected according to He (2020). n/a: not applicable; n.d.: no data.
Sources: State Council (2006a, 2011b, 2016c, 2020, 2021), NDRC (2021), and NEA (2016b).

Targets Disaggregation by Region and Sector

Administrative disaggregation is integral to the planning system in China. Its application in the field of *jieneng jianpai* was introduced by the first CWP for *jieneng jianpai* in 2007 (State Council 2007b). The CWP was issued one year and a half after the beginning of the 11th FYP. The introductory remarks noted that, in the previous year (i.e., 2006), the country as a whole failed to meet the targets related to *jieneng jianpai* as set in the beginning of the year, and that in the first quarter of 2007, six energy-intensive and pollution-heavy industries were growing too fast (6.6 per cent faster than previous year). SNGs, various ministries and centrally administered SOEs were given less than a month to propose countermeasures to get back on track. Meanwhile, the CWP stressed the need to disaggregate *jieneng jianpai* targets and tasks to different munici-palities/prefectures, counties and key enterprises and to enforce the responsi-bility system of SNGs to ensure the fulfilment of these targets.

This system of disaggregation has expanded its scope over the years. This is partly in response to the increased number of binding planning targets. Moreover, by the 13th FYP, disaggregation for EC targets was carried out not only for regions but also for key industries and sectors. The sectoral disag-gregation included eight specific industries. It also specified target indicators (numbers indicated in brackets) for construction (3), transport (5), public institutions (2), and end-use equipment (7) (State Council 2016b). On the other hand, the scope of regional disaggregation also expanded to include RE devel-opment to ensure the fulfilment of the planning target for RE development and to reduce RE curtailment. This involved the specification of provincial targets for non-hydro RE electricity consumption (NHREEC) as a percentage of overall electricity consumption by 2020 (NEA 2016) and more recently the minimum ratio of RE electricity consumption (REEC) in total electricity con-sumption for 2020 (NDRC and NEA 2020). As shown in Table 4.2, the energy intensity reduction target varies from 10 to 17 per cent among the provinces, whereas the permitted increase in TEC differs from 8.5 per cent for Shanghai to 34.1 per cent for Hainan. On the other hand, while the provinces have been assigned a target ratio of NHREEC from 5 per cent to 13 per cent, their RE electricity consumption quotas differ greatly from 11.5 per cent to 80.0 per cent, presumably reflecting the provinces' varying endowment and conditions of RE development. These detailed, disaggregated targets show a sophisticated knowledge about the scope of MLCT potential in different provinces. They serve as the basis for the conduct of the Energy Conservation and Emissions Reduction Target Responsibility System (ECERTRS), to be discussed later.

Table 4.2   *Key MLCT targets by region during the 13th FYP period*

Region	13th FYP Energy intensity reduction target	Total energy consumption (TEC) (10,000 tce) (2015)	Maximum TEC increase (2016–20) (10,000 tce)	Maximum TEC growth rate (2016–20) (%)	Target ratio of NHREEC (2020) (%)	Minimum ratio of REEC (2020) (%)
Beijing	17	6,853	800	11.7	10	15.5
Tianjin	17	8,260	1,040	12.6	10	14.5
Hebei	17	29,395	3,390	11.5	10	13.0
Shanxi	15	19,384	3,010	15.5	10	17.0
Inner Mongolia	14	18,927	3,570	18.9	13	18.0
Liaoning	15	21,667	3,550	16.4	13	15.0
Jilin	15	8,142	1,360	16.7	13	24.0
Heilongjiang	15	12,126	1,880	15.5	13	22.0
Shanghai	17	11,387	970	8.5	5	32.5
Jiangsu	17	30,235	3,480	11.5	7	14.0
Zhejiang	17	19,610	2,380	12.1	7	17.5
Anhui	16	12,332	1,870	15.2	7	15.0
Fujian	16	12,180	2,320	19.0	7	19.5
Jiangxi	16	8,440	1,510	17.9	5	22.0
Shandong	17	37,945	4,070	10.7	10	11.5
Henan	16	23,161	3,540	15.3	7	17.5
Hubei	16	16,404	2,500	15.2	7	32.5
Hunan	16	15,469	2,380	15.4	7	40.0
Guangdong	17	30,145	3,650	12.1	7	28.5

Region	13th FYP Energy intensity reduction target	Total energy consumption (TEC) (10,000 tce) (2015)	Maximum TEC increase (2016–20) (10,000 tce)	Maximum TEC growth rate (2016–20) (%)	Target ratio of NHREEC (2020) (%)	Minimum ratio of REEC (2020) (%)
Guangxi	14	9,761	1,840	18.9	5	39.5
Hainan	10	1,938	660	34.1	10	13.5
Chongqing	16	8,934	1,660	18.6	5	40.0
Sichuan	16	19,888	3,020	15.2	5	80.0
Guizhou	14	9,948	1,850	18.6	5	30.0
Yunnan	14	10,357	1,940	18.7	10	80.0
Xi Zang	10	–	–	–	13	–
Shaanxi	15	11,716	2,170	18.5	10	17.0
Gansu	14	7,523	1,430	19.0	13	44.5
Qinghai	10	4,134	1,120	27.1	10	63.5
Ningxia	14	5,405	1,500	27.8	13	22.0
Xinjiang	10	15,651	3,540	22.6	13	20.0
China	15	447,317	68,000	15.2	n.d.	n.d.

Source:　State Council (2016b), NEA (2016), and NDRC and NEA (2020).

Special Plans and Medium- and Long-Term Plans

Special (i.e., sectoral) FYPs and plans with longer than five-year time spans have been developed in areas ranging from EC, RE, science and technology to industry and urbanization. These are designed to translate the FYP targets into practices in key sectors, often involving centralized coordination. The Medium and Long-Term Special Plan for Energy Conservation (NDRC 2004) was the earliest of this kind. It proposed energy intensity reduction indicators for 14 classes of products and energy efficiency improvement indicators for 10 classes of products. In 2005, the NDRC announced the 'Ten Key Energy Conservation Projects' with the purpose of saving 0.24 bn tce during the 11th FYP period (NDRC 2005). The Medium- and Long-Term Development Plan of Renewable Energy (NDRC 2007b) sets out four overall objectives for the 15 years from 2005 to 2020, including, among others, the increase of the proportion of RE in energy consumption and the industrialization of RE technology. It also sets specific targets for RE development, aiming to bring its share in primary energy sources to 10 per cent by 2010 (the actual was 8.6 per cent) and 15 per cent by 2020. The 12th FYP on Energy Development sets an anticipatory limit on PEC at 4 bn tce by 2015.

Another example is the Action Program to Address Climate Change in Industry (2012–2020), issued by the Ministry of Industry and Information Technology (MIIT), NDRC and the Ministry of Finance (MoF) (MIIT et al. 2012). It requires that by 2015, carbon intensity should decrease by more than 21 per cent in industry compared to the levels of 2010. It envisions that by 2020, this intensity would be 50 per cent lower than the levels of 2005. It further sets individual targets for nine industries including steel, non-ferrous metal, petrochemical, chemical, building material, machinery, light industry, textile, and electronics, ranging from 18 per cent to 22 per cent. These targets mean that manufacturing activities are expected to undergo the LCT at a faster pace than the wider national economy. The MIIT (2016) issued the Administrative Measures for Industrial Energy Conservation Management in April 2016. These measures call on government departments in charge of industrial development and information technology to promote EC in industries, via the scientific establishment of energy use allowance (EUA) and initial allocation of carbon emissions allowance (CEA), and to work towards the trading of EUA and CEA.

Industrial Policy

China's MLCT has progressed with the help of the state's industrial policy. China started to publish the National Industrial Structure Adjustment Guidance Catalogue from 2005 and revised it in 2011, 2013 and 2019. This catalogue essentially classifies all economic activities into four categories: those that are to be eliminated (in short, 'eliminated'), those that are to be restricted ('restricted'), those that are to be encouraged ('encouraged'), and finally, those that are allowed, which covers everything that is not mentioned in the catalogue. Adopting an industrial policy that supports energy conservation and environmental protection (ECEP) was one of the provisions in the 2007 ECL amendment. The State Council's first CWP for *jieneng jianpai* for implementing the 11th FYP (State Council 2007b) sought to achieve *jieneng jianpai* targets by focusing on the 'eliminated' category. Constituted as one of the 45 proposed measures of the plan, the list of eliminated covered 13 sectors and was designed to realize energy savings of 118 million tce during the five-year period. This work continued throughout the 12th and 13th FYPs. Somewhat surprisingly, evidence shows widespread over-fulfilment in this work. For example, during the 12th period, in every one of the 21 areas identified, the elimination target had been exceeded by 2014 (i.e., a year early), often by large amounts (sometimes by more than 10 times) (see NDRC 2016, Table 3-3). Compensation was provided for eliminated productive capacities, but only if the enterprises involved were operating legally and the capacities were built with proper approval and acceptance (Zhao and Li 2018). State ownership of some of these capacities may have made it easier to meet the targets in this area.

By contrast, those activities that are designated as 'encouraged' have received policy support including financial incentives and investment. The example of electric vehicles (EVs) is illustrative of this. The production and use of EVs has been boosted significantly by three successive central plans related to the nurturing and development of the so-called 'strategic emerging industries' (SEIs) (State Council 2010, 2012, 2016a). All of the three plans identified new energy vehicles (NEVs) and related activities as part of the SEIs. The 12th FYP for these industries (State Council 2012) expected the number of cumulative production and sales of NEVs to increase from 500,000 by 2015 to 5 million by 2020, which represents a tenfold increase over five years. The 13th FYP for SEIs further upped the game by forecasting the GDP from EV-related activities to reach RMB1 trillion by 2020 (State Council 2016a). China's NEVs annual addition and total stock reached 1.2 million and 3.8 million, respectively, by the end of 2019, both of which accounted for more than half of global totals; 1.2 million charging points had been installed by then (State Council 2020).

TARGET RESPONSIBILITY SYSTEMS

While the planning system sets out regional and sectoral targets, it is the target responsibility systems (TRS) that have largely ensured the fulfilment of these targets. With increasing emphasis on the 'rule by law' in the post-reform era, China has combined administrative regulations with law to structure the field of action for different subject groups including government officials. Under this system, officials are held legally responsible for achieving the binding planning targets (Wang 2013). A central element in the MLCT governmentality is the Energy Conservation and Emissions Reduction TRS (ECERTRS), which also includes the Energy Conservation TRS (ECTRS). As discussed in Chapter 3, it was the 'Decision on Strengthening Energy Conservation' by the State Council (2006a) that first introduced the ECTRS and the EC Assessment and Evaluation System (ECAES). The decision requires the NDRC to disaggregate the national binding target of energy intensity reduction across province-level jurisdictions, which are then tasked with disaggregating the targets further down to their municipalities, counties and MEUUs and to carry out a strict ECTRS. It further requires that EC performance indicators be incorporated into the comprehensive evaluation and annual assessment systems of economic and social development in various regions; these shall also become an important criterion in the assessment of the performance of leading corps and leading cadres of the government at all levels for implementing the SOD and be incorporated into the performance evaluation of the managers of large- and medium-sized SOEs. These provisions became formalized in the 2007 amendment of the Energy Conservation Law (ECL).

The ECERTRS, which covers both EC and emissions reduction, was operationalized in November 2007 through the announcement of the 'Implementation Schemes and Methods of Statistical Monitoring and Assessment of Energy Conservation and Emissions Reduction' (the 'Implementation Schemes and Methods' in short) by the State Council (2007c) (also see SCPO 2007). These schemes and methods were jointly developed by the National Bureau of Statistics (NBS), NDRC, NEA, and State Environmental Protection Administration (SEPA). They consist of 'three schemes' (*fangan*) and 'three methods' (*fangfa*), corresponding respectively to the energy intensity reduction target and the major pollutant emissions reduction target in the 11th FYP. The three 'implementation schemes' comprise a statistical index system, a monitoring system and an assessment system for energy intensity reduction, whereas the 'three methods' cover the same three aspects for the COD and SO_2 emissions reduction targets.

The ECERTRS was aimed at provincial governments and the selected 'One Thousand Major Energy-Using Enterprises' (MEUEs) during the 11th FYP period. It included a 'Provincial People's Government Energy Conservation Target Responsibility Evaluation and Assessment Score Sheet' and an 'Energy Conservation Target Responsibility Evaluation and Assessment Scoring Sheet for One Thousand Major Energy-Using Enterprises' (State Council 2007c). Both sheets comprise two level-one indicators, related respectively to the fulfilment of the EC target (40 points) and the adoption of EC measures (60 points). Target fulfilment is treated as a veto indicator, meaning that failing to meet this target would lead to the failure of the overall performance. Meanwhile, the indicator of EC measures for provincial governments consists of nine elements, including economic restructuring, financial inputs for energy conservation and implementation of key projects, development and deployment of energy-saving technologies, management of major energy-using enterprises and sectors, implementation of foundational work in EC, implementation of laws and regulations, energy target disaggregation and implementation, organization of EC work and leadership. Since each element covers from one to five sub-elements, the whole scoreboard monitors a total of 26 specific aspects.

Similarly, the scoring sheet for the 1,000 MEUEs includes six categories of performance indicators (weighting in brackets), covering: amounts of energy saved (40 points); EC organization and leadership (5 points); disaggregation of EC target and implementation (10 points); EC technical progress and implementation of EC upgrading (25 points); implementation of EC laws and regulations (10 points); and EC management and implementation (10 points). Notably, the EC target disaggregation is down to workshops, teams or individuals, whereas performance on EC technical progress and upgrading is assessed in four sub-areas including the enterprises' relative performance among sectoral peers, that is, those who are also on the 1,000 MEUEs list.

A series of assessment steps were also prescribed for assessing energy intensity reduction performance annually (NDRC 2007c). The first step is to establish the annual EC targets according to the energy intensity reduction target specified in the national 11th FYP. Second, by end of March, provincial governments and the relevant MEUEs shall submit self-assessment reports on their EC performance in the previous year, which are then independently verified. Third, performance is evaluated by four grades: grade 1 is over-fulfilment (at or above 95 points); grade 2 is fulfilment (80–94 points); grade 3 signifies basic fulfilment (60–80 point); grade 4 represents unfulfillment (below 60 points). Fourth, the assessment and evaluation results shall be submitted to the State Council by the end of May. Fifth and finally, after approval by the State Council, the results are made public by the NDRC. On the other hand, they are passed on to the cadre management departments to be used as an important input for the comprehensive assessment and evaluation of provincial leading corps and leading cadres. The schemes require that this performance is subjected to accountability and 'one vote veto'. Within China's cadre evaluation system (CES), the 'one vote veto' means that if a leader or department fails to meet a performance target carrying this veto, it would cancel out all other work performance (Edin 2003). It thus represents the most powerful sanction.

Furthermore, the assessment scheme specifies:

1. Provincial governments with Grade 1 or 2 shall be commended and rewarded; for those with Grade 4, their leading cadres are not allowed to participate in annual awards or receive honorary titles. The State Council will suspend the approval of new energy-intensive projects in the region.

2. Those provincial governments with Grade 4 must, within one month of the announcement of the assessment results, write to the State Council with proposed measures of rectification within a time limit. If the rectification is not in place in due time, the supervisory department shall pursue the responsibility of the officials in charge of this work in the region.

3. The NDRC and the provincial government shall issue a circular praising the enterprises achieving Grade 1 or 2 to commend and reward them. Enterprises with Grade 4 shall be notified and criticized. They shall not participate in annual awards or receive honorary titles. Nor shall they be given preferential treatment such as exemption from regulatory inspections by the state. Appraisal and approval of new energy-intensive projects and new industrial land use will be suspended. Moreover, within one month of the announcement of the evaluation results, these enterprises must submit remedial measures to the local or provincial government within a time limit. The evaluation results of the state-owned and state-holding enterprises on the 'One Thousand Enterprises for Energy Conservation' programme shall be considered by the state assets supervision administration at all levels as an important basis for the performance assessment of the persons in charge of the enterprises. These will carry a veto too.

Coupled with the binding targets defined by FYPs and disaggregated through the CWPs, these schemes and scoring sheets have helped establish an integrated planning and implementation system for securing *jieneng jianpai* targets. It covers the setting of quantitative performance targets, statistics-based performance monitoring, assessment and evaluation (MA&E), and powerful sanctions.

The ECERTRS was then applied to carbon emissions control when carbon intensity reduction became a binding planning target in the 12th and the 13th FYPs. Table 4.3 illustrates this system for the 12th FYP in relation to provincial governments. It covers four sub-areas including target fulfilment (50 points), tasks and measures (24 points), foundational works and capacity building (26 points), and others. These areas cover a total of 13 indicators, with the highest weightings going to two indicators: CO_2 intensity reduction target for the year (25 points) and cumulative intensity reductions for the 12th FYP period (25 points). It appears that the carbon mitigation TRS places a stronger emphasis on target fulfilment, compared with the ECTRS for the 11th FYP.

Interestingly, while the energy intensity reduction target for 11th FYP was nearly met, the targets for both energy and carbon intensity reductions for the 12th FYP were more than met (see Table 4.1).

The ECERTRS depends on the nationwide energy statistical system to function. Research suggests that while a combination of numerical, result-oriented targets imposed from above and high-powered incentives for hitting targets can induce gaming behaviours among local officials, the internet-based nationwide electricity use system has minimized falsification of such data (Gao 2015). This also, however, explains why some local authorities resorted to drastic measures such as switching off electricity supply to meet the energy saving target set for the 11th FYP (2006–10) in the last quarter of 2010 (Gao 2015). Similar behaviours occurred in 2014, the last year of the 12th FYP (Zhao and Li 2018). An annual monitoring cycle managed by the NDRC have been introduced to mitigate such '13th hour' reactions.

The ECTRS has also been applied to MEUEs under successive rounds of EC programmes targeting these enterprises. The initial scheme, 'Energy Conservation Action by One Thousand Enterprises' during the 11th FYP, covered a total of 1,008 enterprises in nine energy-intensive industries. As of 2004, these enterprises accounted for 33 per cent of national energy consumption and 47 per cent of industrial energy consumption. These enterprises saved a total of 150 million tce, equivalent to a reduction of 388 million tonnes of CO_2 emissions (Zhao et al. 2018). Moreover, under the influence of this national scheme, SNGs launched similar schemes covering lesser MEUEs. Capitalizing on the early encouraging results, during the 12th FYP, the national government launched the scheme of 'Energy Conservation and Low-Carbon Action of Ten Thousand Enterprises' (NDRC 2011b), eventually covering nearly 17,000 enterprises. The aim was to save a total of 250 million tce. In fact, by 2014, the saving had already reached a total 309 million tce. Official data show that in 2014, out of the 13,328 participating enterprises, 31 per cent achieved over fulfilment, and 51 per cent registered fulfilment. Only 7.1 per cent of the enterprises failed to fulfil their targets, compared with 9.5 per cent for 2012 and 8.4 per cent for 2013 (NDRC 2016). During the 13th FYP period, there was a similar scheme, called 'Actions of Hundred, Thousand, and Ten Thousand Major Energy Using Units'. This combines actions by the top 100 MEUUs (with annul energy consumption exceeding 3 million tce) at the national level, the top 1,000 MEUUs at the provincial level (with annul energy consumption exceeding 0.5 million tce), and the top 10,000 MEUUs at prefecture/city level (with annual energy consumption below 0.5 million tce, but above 10,000 tce) (Zhao and Li 2018). Information is lacking on the results of this scheme.

Table 4.3   Indicators for assessing provincial governments on carbon intensity reduction targets during the 12th FYP

Assessment items	Assessment indicators	Scoring	Scoring bases
I. Targets fulfilment (50 points)	1. CO_2 emission per unit of local GDP reduction target	25	Annual planning targets; fulfilment of the verified local annual reduction target
	2. Accumulated reduction of CO_2 emission per unit of local GDP reduction in the 12th FYP period	25	Accumulated reduction target to be met in the current year; verified accomplished accumulated reduction target
II. Tasks & measures (24 points)	3. Progress with the task of industrial restructuring	4	Results of the assessment made by governing authority in the same period; or changes in the GDP contribution of the tertiary industry added value against the previous year
	4. Progress with the task of energy conservation and energy efficiency improvement	4	Results of the assessment made by governing authority in the same period
	5. Progress with the task of energy mix adjustment	4	Results of the assessment made by governing authority in the same period; or changes in the share of hydro, nuclear, wind and solar power in total energy consumption against the previous year or of coal in total energy consumption
	6. Progress with the task of increasing forest carbon sink	4	Results of the assessment made by governing authority in the same period; or up to standard afforestation/tending area increased for the year
	7. Progress with the task of low-carbon pilot and demonstration project development	8	Official documents On site verification

Assessment items	Assessment indicators	Scoring	Scoring bases
III. Foundational works & capacity building (26 points)	8. Target disaggregation and assessment by municipalities or by industrial sectors	4	Official documents. On site verification
	9. GHG emission statistical accounting system building and inventory preparation	6	Official documents. On site verification
	10. Performance of the low-carbon product standards, labelling and certification system	4	Official documents. On site verification
	11. Financial support	6	Official documents. On site verification
	12. Overall management and public participation	6	Official documents. On site verification
IV. Others* (6 points)	Institutional innovation	6	Official documents. On site verification

Note: *These scores are for reference only. The table layout has been adjusted for accuracy according to the Chinese version.
Source: NDRC (2016, Table 5-5, p. 111).

BUREAUCRATIC RESTRUCTURING

To promote MLCT, the Chinese government has also resorted to major re-organization of its large bureaucracy by taking advantage of various rounds of administrative restructuring. As a context, the State Council representing the Chinese central government oversees some 30 central government agencies known as ministries, commissions, administrations, and offices. From time to time, a Leading Group is set up by the State Council to mark the importance of an issue and to coordinate policy actions around it. Such groups are headed either by a vice premier, or by the premier when the issue is a top national priority (Qi et al. 2008). Of all the central agencies, the NDRC acts as the over-arching coordinating body for economic policy making. In this organizational landscape, the establishment of the 'National Leading Group on Addressing Climate Change, Energy Conservation and Emissions Reduction' (hereafter referred to as the Leading Group) in 2007 represented a major milestone. Crucially, it was headed by Premier Wen Jiabao, signifying the importance of the issue. It replaced the National Coordination Group on Climate Change Strategy, which was established in 2003 and led by a vice premier (Qi et al. 2008). The Renewable Energy Law (REL) provides for the establishment of an energy supervisory body under the State Council to set medium- and long-term RE development targets, and for RE to become a priority area to be incorporated into national science and technology development plans and high-technology industry development plans.

During the administrative restructuring in 2008, three important changes were adopted. First, the number of member agencies of the Leading Group was increased from 18 to 20. Second, the all-important NDRC was vested with the task of overseeing the work on climate change. A General Affairs Office of the Leading Group was set up inside the NDRC. Third and finally, a dedicated department, the Department of Climate Change, was established in the NDRC to take charge of the overall coordination of climate change-related work throughout the country (NDRC 2012). Since the NDRC had taken over the supervisory role over energy policy and incorporated the General Administration of Energy in 2003, this arrangement placed the NDRC at the heart of managing the work around both energy and MLCT. Further changes were made. In 2013, the number of the Leading Group membership, now chaired by the current premier, Li Keqiang, increased from 20 to 26 ministries/departments, with seven new additions. In 2015, the National Centre for Climate Change Strategy and International Cooperation (NCSC) was established under the NDRC. Its main responsibilities include the organization of research projects on policies, regulations and planning for addressing climate change nationwide (NDRC 2016).

Moreover, as required by the central government, provincial governments established their own Leading Groups and supporting structures in a similar pattern. Provincial Leading Groups are chaired by their top administrative leaders, with memberships incuding relevant departments and their secretariats in their Development and Reform Commissions (DRC). Research by Qi et al (2008) shows that local governments' attitude towards climate change started to change sharply in mid-2007, and that such leading groups were established in 26 (out of 31) provincial-level jurisdictions between August 2006 and October 2007. Nevertheless, by the end of 2016, only 11 provinces had each set up a dedicated Division of Climate Change within the administrative body that supervises climate change work (MEE 2018a). This may suggest that, while the climate change and LCT issue is firmly placed on the agenda of the central government, the situation is somewhat uneven at the sub-national level. Nevertheless, the bureaucratic restructuring has made the NDRC and its sub-national counterparts throughout the country the 'power stores' and 'power generators' for MLCT, contributing to the making of a technological governing system.

The restructuring of the State Council in 2018 led to the establishment of a 'Ministry of Ecology and Environment' (MEE), which has taken over not only the responsibilities of the MEP, but also the responsibility for overseeing the work on climate change and emissions reduction from the NDRC. Effective from October 2018, this change was designed to strengthen the coordination of climate change and environmental pollution prevention and control, and to enhance the integrity of ecological environment protection (MEE 2018b). However, because of this restructuring, the function of overseeing climate change policy has been taken out of the NDRC, although it continues to oversee energy policy. The effect of this change is unclear for the time being.

RE-ORDERED PERFORMANCE CRITERIA AND STRENGTHENED ACCOUNTABILITY FOR GOVERNMENT OFFICIALS

As shown earlier in this chapter, CES and TRS are key elements in carbon governmentality in China. In their 'Suggestions on Accelerating the Construction of Ecological Progress' ('the Suggestions' hereafter), the CCPCC and State Council (2015b) called for the perfection of the performance evaluation system by incorporating indicators covering resource depletion, environmental damage and ecological benefits, and by increasing their weightings. More importantly, they proposed the establishment of a responsibility system for leading cadres for the construction of Ecological Progress and improvement to the responsibility assessment and accountability system for *jieneng jianpai* targets during officials' tenure of office. Increased emphasis is placed on individual accountability, and indeed lifelong accountability. To formalize these proposals, the CCPCC and the State Council (CCPCCGAO and SCGAO 2015) issued the 'Measures for Investigating the Responsibility of Leading Party and Government Cadres for Eco-Environmental Damage (Trial Implementation)' in August 2015. This was followed by the 'Assessment Methods for Ecological Progress Construction Objectives' from the same sources (CCPCCGAO and SCGAO 2016) in December 2016. Finally, a Green Development Index (GDI) system and an Ecological Civilization Construction Target Assessment System were introduced in December 2016 (NDRC et al. 2016).

Applied again to both regional party and government leaders, the Ecological Civilization Construction Target Assessment System combines annual assessment with five-yearly evaluation. The annual assessment is based on the compilation of the GDI for each regional jurisdiction. The Index is the weighted sum of 55 level-two indices under six classes of level-one indicators (see Table 4.4). Different weightings are attached to three kinds of indicator. While the binding resource and environment indicators indicated in the 13th FYP are given a weighting of 2.75, other monitoring and evaluative indicators identified either in the FYP or in the 'Suggestions' carry a weighting of 1.83. Finally other important green development monitoring and assessment indicators have a weighting of 0.92. It is worth noting that, in the GDI system, growth-related indicators now carry only 9.2 per cent of the overall weighting in the index, compared with a combined weighting of 10.8 per cent for the four LCT-related binding indicators in the 13th FYP. GDP per capita growth, a standard measurement of economic growth, only carries a weighting of 1.83 per cent, the same as attached to disposable income per capita and the share of tertiary sector GDP. Clearly performance priorities have been re-ordered.

COMPILATION OF GHG INVENTORIES

There is much emphasis on quantifying emissions and mitigation pathways. At the national level, China has published its national GHG inventory up to 2014 (MEE 2018b). This shows that, as of 2014, energy, industrial processes, agriculture and waste disposal accounted for 77.7 per cent, 14.0 per cent, 6.7 per cent and 1.6 per cent of the total GHG emissions (excluding land use, land-use changes and forestry) respectively. Such emission structure has informed policy focus on energy and industry.

In September 2010, the Department of Climate Change, NDRC, issued a *Circular on Starting the Preparation of Provincial Greenhouse Gas Inventories* (No. 2350). This marked the beginning of a nationwide programme to develop provincial GHG inventories (NDRC 2017a). In May 2011, the NDRC (2011a) published the 'Guidelines for the Compilation of Provincial Greenhouse Gas Inventories (Trial)'. The compilation of provincial GHG

Table 4.4　　The Green Development Index system

Level-1 indicator	Weighting (%)	No. of level-2 indicators	No. of 13th FYP binding indicators	Binding indicators
Resource use	29.3	14	6	1. Reduction of energy consumption per unit of GDP 2. Reduction of CO_2 3. Emissions per unit of GDP 4. Share of non-fossil fuel energy in primary energy consumption 5. Water use per unit of GDP 6. Protected cultivated land 7. Land use for new construction
Environmental governance	16.5	8	4	1. Reduction of total emissions of chemical oxygen demand (COD) 2. Reduction of total ammonia nitrogen emissions 3. Reduction of SO_2 emissions 4. Reduction of nitrogen oxide emissions
Environmental quality	19.3	10	4	1. Ratio of days with excellent- and good-quality air for cities of and above municipality status 2. Reduction of PM2.5 concentration for cities of and above municipality status 3. Share of surface water at or better than class 3 4. Share of class 5 surface water
Ecological protection	16.5	10	2	1. Forest area coverage 2. Cubic quantity of forest
Growth quality	9.2	5	0	None
Green living	9.2	8	0	None

Source:　　NDRC et al. (2016).

inventories for 2005 and 2010 was completed by the end of 2014. Furthermore, in January 2015, the NDRC initiated the compilation of provincial GHG inventories for 2012 and 2014 (Li et al. 2017). To aid this work, from October 2013 to July 2015, the central authorities issued individual guidelines for GHG accounting and reporting across 24 industries (NDRC 2016).

To explore effective ways of doing this, six provinces, including Zhejiang province, and one municipality (Tianjin) were designated as pilots for this work in 2010. By August 2017, the seven pilot regions had completed their 2005 inventories covering five areas including energy activities, industrial production processes, agriculture, land-use change and forestry, and waste treatment. The same work was still ongoing in the other 24 provincial jurisdictions as of 2017 (NDRC 2017a). As a consequence of these efforts, Chinese SNGs have started to gain an understanding of where the greatest mitigation potential lies and how to develop roadmaps for achieving their MLCT objectives (see Chapter 6).

SUB-NATIONAL PILOTING AND EXPERIMENTATION

In China, pilots are often used to test and fine-tune new policies or measures before their wider introduction across the country. There are usually specific incentives for encouraging early birds before a certain policy or measure is rolled out. Heilmann (2008) makes a distinction between general piloting and transformative experimentation in China. While the former is characterized by narrowly defined trial measures and preselected target groups, the latter is characterized by a search for creative options implementing selected tasks. Experimentation can take three different forms: experimental regulation, experimental pilots (*shidian* in Chinese), and experimental zones.

Key experimental pilot programmes for LCT include low-carbon city pilots (LCCP), carbon emission trading (CET) pilots, low-carbon industrial park pilots, low-carbon community pilots, low-carbon transport system pilots, energy use allowance (EUA) trading pilots and so on (see Table 4.5 for an incomplete summary). The LCCP programme, covering three batches, has had the greatest reach and impact of all pilot schemes. The 42 pilot cities and provinces belonging to the first two batches alone accounted for 57 per cent of national GDP, 42 per cent of population, and 59 per cent of energy consumption as of 2015 (Li et al. 2017, p. 10). By January 2017, a total of 87 sub-national jurisdictions, including 78 cities, had been designated as low-carbon pilots in three batches. Many low-carbon pilots, including all the 45 pilots of the third batch, have announced their target emission peak years, 17 of which fall within the 13th FYP (2016–20).[1] Almost two-thirds of the cities belonging to the first and second batches set their carbon intensity reduction targets higher than 45 per cent (i.e., the upper end of China's Cancun Pledge). The cities have implemented policies in four main areas: energy efficiency improvement, RE development; sector structure adjustment, and carbon sequestration capacity increase (forestry) (Wang et al. 2015).

Table 4.5   A selection of state-level low-carbon pilot schemes in China

Pilot scheme	Launch date	Responsible government department	No. of pilots
'Ten cities, thousand vehicles' EV demonstration project	Jan 2009	MST, MOF, NDRC & MIIT	25 cities (over 3 years)
1st batch of low-carbon province and city pilots	Jul 2010	NDRC	8 cities and 5 provinces
Carbon emission trading pilots	Jun 2011	NDRC	7 cities and provinces
Green and low-carbon small town pilots	2011	MoF, MoHURD & NDRC	7 small towns
Low-carbon transport system pilots	2011 & 2012	Ministry of Transport	26 cities
2nd batch of low-carbon city and province pilots	Dec 2012	NDRC	27 cities, 1 region and 1 province (Hainan)
1st batch of low-carbon industrial park pilots	May 2014	NDRC & MIIT	55 cities identified
Low-carbon community pilots	2014	NDRC	Target: 1,000
Low-carbon cities (towns)	Aug 2015	NDRC	8
Energy use allowance trading pilots	Jul 2016	NDRC	4 provinces
3rd batch low-carbon city pilots	Feb 2017	NDRC	41 cities, 1 urban district, 3 counties
Green finance reform and innovation pilots	Jun 2017	State Council	5 provinces

Note: NDRC = National Development and Reform Commission; MoF = Ministry of Finance; MoHURD = Ministry of Housing and Urban-Rural Development; MIIT = Ministry of Industry and Information Technology; MST = Ministry of Science and Technology.
Sources: https://www.ndrc.gov.cn; Zhang (2015), Zhao and Li (2018), Wang and Zadek (2017).

Interestingly, financial incentives have not been used as a key instrument in the LCCP programme. Instead, the focus has been on exploring market-based

---

[1]   These included 12 pilots among the first and second batches (GLCDTTP 2016) and 5 from the third batch (NDRC 2017b).

mechanisms. Indeed, Jiang Zhao Li, deputy director of the Department of Climate Change, NDRC, suggests that the programme has broken new ground in China's tradition of piloting because it is piloting from bottom-up and that the key is for cities to identify low-carbon pathways appropriate for their local conditions (21st CBH 2016). He argues that this reflects the centre's emphasis since the 18th Party Congress (2012) on promoting socio-economic development through marketization and legislation, rather than administrative incentives.

MANDATORY ENERGY EFFICIENCY STANDARDS AND RENEWABLE ENERGY PERFORMANCE REQUIREMENTS

Setting MEES and introducing compulsory energy efficiency labelling (EEL) have been two important tools for protecting the environment and saving energy since the 1990s in China. Their importance has increased since the introduction of an EEL management system mandated by the 2007 amendment of ECL. This obliges ECAE makers and suppliers to state the norm of the ECAE's energy efficiency levels through labelling. Moreover, the amendment also mandates MEES for ECAE and energy consumption quotas (ECQ) for energy-intensive production processes. Appliances and equipment that fail to meet the MEES are to be eliminated. Applications for fixed asset investment (FAI) projects that fail to meet the ECQ shall be rejected. The amendment also spells out punishment for violation, including business closure, investigation and fines. Since 1981, more than 400 EC standards have been announced, about half of which are at national level (Zhao and Li 2018). During the 12th FYP period alone, 221 EC national standards were issued, covering industry, energy, building, transport, and public sector institutions. By the end of 2015, EEL had covered 33 categories of end-use appliances involving more than 10,000 registered enterprises and more than 930,000 specific products. In the ten years after its introduction, the EEL system had reportedly saved a cumulative amount of energy in excess of 441.9 billion kWh (NDRC 2016). IEA (2018b, p. 147) reports that, with around 60 per cent of final energy use covered by MEES policies in 2017, China had the highest coverage globally. China has also introduced an energy utility obligation programme in 2009, although its strength is low (IEA 2018b, p. 41). By 2020, power suppliers were required to source at least 9 per cent of their power supplies from non-hydro RE sources or meet that requirement with renewable energy green certificates (NEA 2016).

MARKET-BASED MECHANISMS

A variety of experiments with the use of market-based mechanisms has been carried out. These include market-assisted vehicle licence allocation, energy pricing reforms, and trading of carbon emissions allowances (CEAs), certified reductions, and EUAs.

Market-Assisted Allocation of Vehicle Licences

Several Chinese cities have experimented with a market-based approach in allocating vehicle licence quotas. By December of 2014, seven Chinese cities had implemented licence plate restriction policies, either by auction or lottery, or a combination of both. While Shanghai used auctions, Beijing and Guiyang used a lottery. The other four cities used a combination of both auction and lottery (Xiao et al. 2017). The experiences of Shanghai and Beijing are illuminating. Shanghai was the first Chinese city to introduce a vehicle licence auction, as early as 1992 (arguably back to 1986). Introduced initially to curtail traffic congestion, it was also used to support the local car-making industry between 1998 and 2000 (Heyanyueche 2018).

The system in Shanghai saw the average price of a licence increase from RMB 27,040 in October 2002 to RMB 90,687 in 2017, while the success ratio fell from 68.65 per cent to 4.46 per cent (Zhihu 2018). Moreover, auction in Shanghai raised RMB 5 billion in 2011 and an estimated RMB 21 billion in 2012 for the public coffer (Yang et al. 2014). Research suggests that the high-priced licences in Shanghai are effective in controlling the sales of passenger vehicles, although the positive effect of vehicle control on the environment is partially offset by lowered fleet energy efficiency because the restriction screens out lower-income buyers, who tend to buy smaller and more fuel-efficient cars, compared with higher-income buyers (Xiao and Zhou 2013). Specifically, research shows that the scheme in Shanghai lowered car sales by 48.4 per cent in June 2010, and reduced fuel consumption by 47.67 per cent. The difference (1.42 per cent) represents an increase in fuel consumption due to lower fuel efficiency. Critics complained that this system makes it harder for car manufacturers to meet other central government policies like the average corporate fuel consumption quota, as specified in the 'Passenger Vehicle Corporate Average Fuel Consumption Calculation' effective since 2012. The latter requires that the average corporate fuel consumption coefficient must be no more than 6.9 litres/100 km by 2015 (or equivalently, no less than 14.49 km/litre).

Beijing became the first Chinese city to introduce a lottery-style system in 2011. Research has found that growth in new vehicle registration was sharply curtailed in Beijing because of the lottery. New car registration was cut by over 75 per cent between 2010 and 2011 (Yang et al. 2014). However, it was estimated that the effect on fuel consumption by cars is limited for two reasons. First, cars are driven more intensively when congestion improves. Second, similar to Shanghai's experience, buyers in a restricted licence plate regime tend to concentrate their purchases on larger, less fuel-efficient cars. As a result, although the lottery will have reduced the number of vehicles by 11 per cent from a no-policy scenario by 2020, this policy will reduce fuel consumption by only 1 per cent.

Energy Pricing Reforms

The 1980s saw a trend towards price liberalization for commodities, but not for most forms of energy. The pricing treatments for end users of electricity and oil products and the producers of coal, crude oil and oil products were highly uneven. While the former were subjected to tight control because of equity and social stability concerns, the latter were put under a degree of market pricing, depending on whether the production was planned or above-plan (Andrews-Speed 2012). Much has changed since then. The details of China's energy pricing reforms have been exhaustively explained by Zhang (2018) and will therefore not be repeated here.

It suffices to state that, while earlier periods focused on the price liberalization of energy, the government started to exploit the potential of EC through the lever of energy pricing from the 11th FYP (Zhao and Li 2018, p. 36). The measures include differential pricing, punitive electricity prices, and tiered pricing for electricity. Differential electricity pricing was first introduced in June 2004, and was designed to apply differential electricity prices to four types of enterprise defined by the state's industrial policy in six high energy-consuming industries such as electrolytic aluminium. To improve implementation, the NDRC stepped this up by banning SNGs from subsidizing energy-intensive enterprises, expanding the scope of this policy from six to eight industries, and strengthening the price differences (electricity prices for the 'eliminated' type were set at 50 per cent above those for the high energy-consuming sectors) (SCGAO 2006).

From 2010, the government started to apply punitive electricity prices for enterprises whose energy consumption exceeds standards set by the central or/and local government. If an enterprise's energy consumption exceeds the standards by 100 per cent, the enterprise is charged the same kind of punitive prices as applied to the 'eliminated' type of enterprise (i.e., 50 per cent more). Meanwhile, since 2004, the government has applied power price premiums for power suppliers that have completed desulphurization and denitrification. This seems to have been effective. By 2011, the portion of coal-fired units fitted with a flue-gas desulphurization facility rose to 90 per cent of the total installed thermal capacities, compared with 13.5 per cent in 2005 (Zhang 2018, p. 515).

Since July 2013, a two-tiered natural gas pricing system for residential users has been introduced. On the one hand, a lower price is set for the baseline 2012 consumption volume, with city gate prices capped, whereas a higher price is set for any volumes above the 2012 consumption level. The post-2015 pricing mechanism for gas sets three price bands for three tiers of consumption for household users in the region. Depending on whether the consumption falls within 80 per cent, 95 per cent, or above 95 per cent of average monthly consumption volumes for household users, the price would vary. Consumption at the 2nd and 3rd tiers is charged at 120 per cent and 150 per cent of the first-tier price, respectively (Zhang 2018). In 2011, the NDRC has introduced tiered electricity prices with regards to residential electricity consumption in three percentiles of the region: below 80 per cent; 80–95 per cent; above 95 per cent. The price for the 3rd tier is 1.5 times the price of the second tier (Zhao and Li 2018).

CET and EUA Trading

From the 12th FYP onward, China's MLCT governing system gradually shifted its emphasis from a mix of administrative and legal means to economic means (Zhao et al. 2018). Experiments have been carried out on CET and EUA trading. However, the results are somewhat limited so far.

Carbon emissions trading

NDRC issued in 2011 the 'Circular on Carrying out Carbon Emission Trading Pilots' and chose Shanghai, Beijing, Guangdong (province), Shenzhen, Tianjin, Hubei (province), and Chongqing as the pilots. The pilots were given little guidance from the NDRC about what to do. Instead they were required to develop a governing structure for such trading, to clarify the rules of the game, to calculate and determine local GHG emission caps, to formulate local GHG emissions quota allocation schemes, to put in place supervision systems and registration systems for local CET, and to establish trading platforms (NDRC 2016).

These pilot schemes operated from June 2013 to June 2014. Together they included a total of 2,667 MEUEs (Liu et al. 2015). Their emissions as a percentage of their region's total emissions varied between 33 per cent and 60 per cent. The opening price of CO_2 ranged from RMB 28/ton to RMB 61/ton (USD 4.57 to USD 9.95). However, these pilots suffer from two main problems. First, due to low trade volumes, the prices traded appeared to be not mainly determined by the market, but by mutual agreements of trading partners. Second, the markets are localized (Liu et al. 2015). In other words, there are no exchanges between these pilot schemes. As a result, the cumulative effects are still limited. By the end of August 2020, the cumulative volume and value of transactions nationwide was 406 million tonnes of CO_2 and RMB 9.28 billion, respectively (An 2021). The average carbon prices of the seven Chinese pilots were only one-third of all other carbon markets in early 2018 (Weng et al. 2018, p. 62). Thus, despite the positive learning and awareness-raising effects, these pilots have not had any material impact on mitigation in China so far, as both carbon prices at the pilot schemes and the amounts of emissions traded are still low.

Preparatory work for a unified national ETS has been ongoing since December 2017, when the Scheme for Constructing the National Carbon Emission Trading Market (Power Generation Sector) was issued. The plan was to start active trading in 2020. It was expected that the turnover could be as high as 3 gigatons, dwarfing that of the European ETS at 1.7 gigatons (SCPO 2018). In December 2020, the MEE initiated the implementation plan for setting and allocating the emission allowances among 2,225 key emitters in the industry, for formal trading to commence from 2021 (MEE 2020).

EUA trading

The impetus of EUA trading comes from the official policy to impose a cap on the TEC in the 13th FYP for Energy Development, as indicated earlier. EUA trading was first formally proposed in the 'Integrated Reform Plan for Promoting Ecological Progress' in 2015. The 'Energy Production and Consumption Revolution Strategy (2016–2030)' (EPCRS) (NDRC and NEA 2016a) also endorses it.

EUA trading is like CEA trading in the sense that the system sets a cap on TEC and then allocates this amount as quotas among energy users (either freely or by auction) such as regions and enterprises. This allows the application of the disaggregation method that the administrative system is accustomed to. The government's intention was that during the 13th FYP period, a system of primary EUA allocation would be established and trading of EUA would commence (CNR 2016). Four provinces were instructed to experiment with the system of paid use and trading of EUA in 2016. They have been given one year (2016) to conduct top-level design and preparation, two years to initiate the experiment, one year (2019) for drawing lessons and developing replicable measures and systems, and a final year (2020) for nationwide evaluation (NDRC 2016).

While some researchers consider EUA trading a Chinese innovation, Duan et al. (2018) are of the view that the parallel existence of carbon emissions trading and EUA trading is a mark of poor coordination between two divisions within the NDRC who are responsible for drafting relevant regulations for energy conservation and carbon emissions trading, respectively. In any case, this has not had much practical impact so far.

FINANCIAL INCENTIVES AND INVESTMENT

Unsurprisingly, incentives and investment through public finance constitute a key governing technique. As Chapter 5 will show, finance has been provided through both increased allocation from public budgets, the banking industry and capital markets. In particular, the EC special funds within the General Public Budget (GPB) and the Renewable Energy Development Special Fund (REDSF), which is placed within the State Funds Budget, have played a key role. By 2010, the special funds *for jieneng jianpai* and new energy already had an annual allocation of RMB 100 billion (Green Sohu 2010). Furthermore, special budgetary funds for transport and circular economy were set up respectively in 2011 and 2012. The rules of the transport special fund were: (1) for projects whose emissions reduction can be quantified, prize money is awarded (up to RMB 600 per tce or up to RMB 2,000 per ton of standard oil for substitute fuels); (2) for projects whose emissions reduction cannot be quantified, subsidies up to 20 per cent of equipment purchase or construction costs can be awarded. However, the subsidy for an individual project should not exceed RMB 10 million (MoF and MoT 2011). Overall, financial incentives and preferential debt mechanisms have played a major role in the installation of RE generation capacities and the production and purchase of EVs.

Subsidies for RE

China's RE investment is legendary. Its share of global RE investment was more than one-third in 2015 and 45 per cent in 2017 (Qi 2018, p. 5). This phenomenal performance is partially based on its feed-in tariffs (FiT) system, introduced initially by the REL in 2006 and consolidated by the 2009 REL amendment. The REL stipulates that cost differences between the purchase of RE-sourced electricity and the average cost of generating electricity through conventional energy sources by grid companies can be added to electricity distribution prices and that a budgetary REDSF shall be established. The chief objective of the REDSF is to provide support for RE research, the development of RE standards and RE demonstration projects, and localized production of RE equipment. Moreover, grid companies shall be held liable and fined for failing to purchase all RE-sourced electricity. There should be preferential treatment for RE in bank lending and taxation. Furthermore, the 2009 amendment of the REL commits that 'The state implements a fully guaranteed purchase system for electricity generated from renewable energy sources.' It stipulates that relevant departments under the State Council shall determine the proportion of RE-sourced electricity to total electricity supplies in a planning period. It also clarifies that the cost differences shall be compensated by a nationwide RE Levy on sales of electricity; and that the REDSF shall be funded by annual budgetary allocation and revenues generated from the RE Levy. Moreover, if a grid company cannot recover its electricity costs through electricity sales, it may apply to the REDSF for subsidies. The REDSF was subsequently established in 2011. After four rounds of adjustment, the RE Levy has stabilized at RMB 0.019 per kwh since 2016 for energy users in the secondary and tertiary industries (Everbright Securities 2019).

The post-2011 period has been a 'golden age' for RE development in China (Everbright Securities 2019). Between 2012 and 2018, the RE Levy raised a total of RMB 338.478 billion, which provided about 85 per cent of the subsidies dispensed by the REDSF.[2] In comparison, RE subsidies for the period from January 2006 to April 2011 amounted to only RMB 32.178 billion. State subsidies supported the installation of a total of 177.77 GW of RE generation capacity between June 2012 and March 2019 (Everbright Securities 2019). The government has also made use of market-based mechanisms for discovering RE prices and bringing down costs. It initially used concession bidding (2003–10). Between 2006 and 2012, however, the main instrument was a regionally based system where regional governments were responsible for collecting the RE levies and distributing the revenue, whereas the central government allocated subsidy quotas. This caused large financial imbalances between those provinces that collected the revenues and those that needed the subsidies. Thus finally, since 2012, the system has combined the nationwide benchmark FiT, centralized revenue allocation and set up a subsidy scheme that RE producers can apply to, although provinces remain responsible for collecting the revenues (Everbright Securities 2019). As a result of the massive investment and the market mechanisms, significant progress in RE has been made. By 2015, the share of non-fossil fuel power generation installations had reached 34.3 per cent, 7.7 per cent higher than in 2010; this ratio further rose to 36 per cent by 2016 (Li et al. 2017). China's RE generating capacities from wind and PV grew from 2,070 MW and 3,500 MW in 2006 to 185 GW and 175 GW in 2018, respectively. Finally, according to the International Renewable Energy Agency (IRENA) (2019), China's levelized cost of energy (LCOE) for wind decreased by 73 per cent between 1996 and 2018 to reach $0.047/kWh (the lowest in the world), whereas the LCOE for PV decreased by 77 per cent between 2010 and 2018 to reach $0.066/kWh. The latter is lower than those in

---

[2]   This assumes that allocation from the GPB was maintained at 15 per cent. The actual figures for 2015 and 2016 were 15.7 per cent and 16.1 per cent respectively according to Everbright Securities (2019).

Japan, the USA, Britain, Germany and France, but higher than those in Italy and India. From May 2019, China's policy emphasis has shifted to normalize price parity between conventional energy and RE and the complete marketization of RE.

Nevertheless, there have been some quite serious problems. On the one hand, there is a widening gap between entitled subsidies and actually paid subsidies. This is partly because the system never managed to collect more than 70 per cent of what could be collected from the RE Levy (Everbright Securities 2019). On the other hand, the combination of guaranteed access to power grids and protected prices for wind energy from 2009 to 2015 led to a huge surplus in installed capacities, especially in northern and western China, where wind resources are rich and electricity demand is limited (Zhang 2018). A similar problem existed with solar power installations. This was further aggravated by slower economic growth near the end of the 12th FYP: annual electricity demand growth fell to 3.4 per cent and 0.5 per cent in 2014 and 2015, respectively. This combination led to a high level of abandonment of RE generation capacities. For wind power, the curtailment ratio rose from 10 per cent in 2010 to 17 per cent in 2016. The total amount of abandoned energy, encompassing wind, solar, hydro and nuclear power, was conservatively estimated at 150 billion kWh in 2016 (Li et al. 2017).

The high RE curtailment levels led to the launch of a three-year 'Clean Energy Consumption Action Plan (2018–2020)' by the NDRC and the NEA (2018). This action plan specified minimum annual clean energy consumption targets for key RE producing provinces between 2018 and 2020, aiming to bring the curtailment rates for wind, PV and hydro power down to 5 per cent. The plan made meeting clean energy consumption targets a prerequisite for adding new clean energy capacities and banned areas with wind and PV curtailment from building beyond what is permitted by the 13th FYP. On the other hand, the 'Renewable Energy Consumption Guarantee Mechanism' (NDRC and NEA 2019) issued minimum and incentivized RE consumption quotas for all provinces (except Xi Zang). It also introduced an annual monitoring mechanism and offered the exclusion of RE consumption from TEC control – part of the 'double-control' mechanism – as an incentive for consuming RE. These measures appear to have worked. By 2020, curtailment ratios for wind, PV and hydro generating installations had fallen to 4 per cent, 2 per cent, and 4 per cent respectively (NEA 2020). This dramatic turnaround once again shows that a mixture of administrative, financial and market-based mechanisms can be effective in ensuring MLCT development.

Subsidies and Investment for EVs

The central government started to encourage the use of EVs in 2009. The 'Ten Cities, One Thousand Vehicles Energy Conservation and Alternative Energy Demonstration Programme' was initiated jointly by four ministries in January 2009. It aimed to launch 1,000 NEVs (mainly for use in public transport and services, taxis, postal service, and the like) in 10 additional cities every year over a three-year period (2010–12) and to increase plug-in electric vehicle (PEV) sales to 10 per cent of automotive sales nationwide by 2012. Started with 13 cities, the project eventually involved 25 cities over the three-year period. Twenty-four models of NEVs (in two batches) were eligible for government subsidies (Baidu 2021b). The 12th FYP for Strategic Emerging Industry Development (State Council 2012) introduced a piloting scheme to subsidize the purchase of NEVs by private individuals. The latter has had a major impact on the uptake of NEVs.

The market was slow to develop in the beginning. In 2013, only 17,600 PEVs (mostly buses and utility trucks) were sold, accounting for less than 0.1 per cent of total civilian vehicle sales (Wan et al. 2015). However, upon entering the 13th FYP period, both production and sales have picked up speed. In 2017, the output of NEVs in China was 794,000, an increase of 53.8 per cent over the previous year (CHYXX 2018). According to data from Bloomberg, the sales number of alternative energy passenger vehicles in China has increased from 233.7K (January 2015–16), 418K (January 2016–17), 762.7K (January 2017–18), 1.3m (January 2018–19) to 1.5m (January–September 2019), with corresponding increases in their ratio to the sales of all passenger vehicles from 1.1 per cent to 1.7 per cent, 3.1 per cent, 5.6 per cent, and 6.8 per cent (Bullard 2019). The same source highlights several other developments. First, accompanying the increased use of NEVs, China's demand for gasoline has slowed down since 2016, whereas its demand for diesel peaked in 2015. Second, Chinese EV makers claimed half of the top 10 places globally according to estimated 2018 revenues from electric passenger vehicle sales. Third, in terms of the percentage of estimated revenues from these sales as a proportion of total sales, Chinese carmakers claimed 9 out of the top 10 places.[3] In particular, Chery and BYD generated more than 40 per cent of their sales revenue from these sales. Moreover, the EVs market has transformed. By May 2020, vehicle insurance data show that 71.4 per cent of the owners of newly registered EV were individual users (Deng 2020).

---

[3]   The only non-Chinese carmaker that made the cut was Mitsubishi. Tesla was excluded.

Research suggests that three key factors are behind this impressive turn-around: government subsidies, free licences and time saving. The second and third elements are related to the fact that while the allocation of licences by auction or lottery in some Chinese cities has made it both expensive and time-consuming to get vehicle licences, EVs come with free licence plates. On the other hand, subsidies have certainly played a decisive role in propelling the industry to its current scale. Wan et al. (2015) reported that Shanghai provided cash subsidies of about USD 6,540 (RMB 40,000) plus free private vehicle registration plates to buyers of selected EVs. Given that such a plate was worth about USD 12,000 (RMB 74,000) in April 2014, the total subsidies were worth over USD 27,600 (RMB 171,000). However, they pointed to four main problems affecting the industry: protectionism by local governments, technological uncertainty, lagging investments in charging infrastructure, and conservative investment behaviour by automakers and battery manufacturers. Typical protectionist measures include exclusion or phased inclusion of non-local EV brands for subsidies offered by local governments. As an example, the authors cited reports that BYD's EVs were initially excluded from Shanghai's subsidy eligibility list. Local governments could also support manufacturing and purchase of EVs through low land-use fees, direct investment, the installation of charging points and the like. Nevertheless, there is some evidence that protectionism may have lessened or is not as serious as claimed. For example, Masiero et al. (2016) show that in 2015, as many as 17 cities offered subsidies for four different EV models produced by BYD. The highest subsidies were for pure electric buses, up to USD 80,100 per unit. Interestingly, there was a high level of parity in the amounts of subsidies offered across cities for the same models.

However, demand-side factors are only part of the story. Two supply-side factors have also played a part. On the one hand, the 'regionally decentralized authoritarianism' (RDA) system is at work. Regional competition is likely to be responsible for the fact that at one point, more than 500 manufacturers were reportedly registered to make EVs in China (Wharton School 2019). On the other hand, competition by EV manufacturers backed by their home SNGs has produced a dynamic industry. For example, having successfully attracted the likes of Tesla to set up a giga-factory in Shanghai, the city has finally caught up with Shenzhen – the home of BYD (standing for 'Building Your Dream'). According to car insurance data, by May 2020, Tesla (China), which started its production in October 2019 in Lingan New City in Shanghai, had emerged as the top EV maker with a single model (Model 3) in the country (Deng 2020). Favourable treatment by banks and the capital market under China's green finance programme is another important factor for the growth of EVs in China, as we will show in Chapter 5.

TRAINING, EDUCATION AND CAPACITY BUILDING

The ECL calls for the strengthening of EC awareness and education and the spread of scientific knowledge about EC, while its 2007 amendment further mandates the incorporation of EC knowledge into national education and training systems. Meanwhile, the REL requires that RE knowledge and technologies be incorporated into the curricula of general and occupational education. A national Energy Conservation Awareness Week was established in 1991. The third day of the Energy Conservation Awareness Week has been set as the national 'Low-Carbon Day' since 2013. Furthermore, as shown earlier, within the ECERTRS, a key assessment element for SNGs and MEUUs is institutional capacity building. As we will see in Chapter 5, much emphasis has been placed on the strengthening of institutional capacity in addressing *jieneng jianpai* among financial institutions too.

The results of such governing efforts are that there seems to be a high awareness level of EC and climate change issues among the Chinese population. A 2016 survey by Harvard Kennedy School of Government found that '75 per cent of all respondents believe that climate change is real and caused by human behavior, and nearly 70 per cent support enacting a nationwide emissions tax; far higher percentages than rates found in the United States' (Cunningham et al. 2020, pp. 11–12). Research shows that 98.1 per cent of urban consumers have a certain understanding of EC labelling (NDRC 2016).

CONCLUSION

Analysis in this chapter has identified an extraordinarily wide range of GTTs deployed by the Chinese central government for promoting *jieneng jianpai*, RE and EVs, covering not only legal and administrative mechanisms, but also market mechanisms to affect the conduct of different actors. Furthermore, the system shows a strong focus on performance optimization by putting emphasis on setting quantitative binding targets, monitoring, assessment and evaluation, and sanctions. Moreover, through the lens of different forms of GTT (see Table 4.6), the analysis shows that these technical means are distributed across juridical, disciplinary and governmental forms of power. On the other hand, it is interesting to note that while juridical and disciplinary powers have been applied to SNGs and officials, SOEs, MEUUs, power suppliers, ECAE manufacturers and suppliers, government power through price- and market-based mechanisms, awareness campaigns and institutional capacity building has been mobilized to induce and incentivize low-carbon behaviours among producers, households and consumers. In the final analysis, however, the tripartite system of ECERTRS, consisting of performance targets for SNGs and MEUUs and their leaders, statistics-based annual and five-yearly performance assessments and evaluation, and sanctions, is by far the most critical in helping meet China's MLCT planning targets.

This analysis also suggests that China's carbon governmentality has adopted some aspects of what Dean (2010) has discussed in the context of 'advanced liberal government' (ALG). This is so in the utilization of two distinct, yet intertwined technologies, namely technologies of agency and technologies of performance. Dean (2010) points out that the former seek to enhance and improve the targeted groups' capacities for participation, agreement and action, whereas the latter render these capacities calculable and comparable so that they might be optimized. He states: 'If the former allow the transmission of flows of information from the bottom, and the formation of more or less durable identities, agencies and wills, the latter make possible the indirect regulation and surveillance of these entities' (Dean 2010, p. 202). The former in turn are further divided into two elements: contractualization and what Barbara Cruikshank (Cruikshank 1993) has called 'technologies of citizenship' including the multiple techniques of self-esteem, empowerment, and consultation and negotiation. These technologies are designed to engage the targeted populations 'as active and free citizens, as informed and responsible consumers, as members of self-managing communities' (Dean 2010, p. 196).

In the case of China, the use of the 'technologies of citizenship' is relatively rare, at least so far. However, contractualization has been pervasive in the management of SNGs, government officials and MEUUs. Nevertheless, what distinguishes the carbon governmentality in China is its application of the technologies of performance among state actors, rather than actors in the civil society as in the ALG. Furthermore, this system capitalizes on a pre-existing CES and a comprehensive nationwide system of informational and monitoring infrastructure. There is little doubt that what we have discovered is an assemblage of significant complexity and diversity, which clearly meets the thresholds that Dean (1996) set for technological government. While it is necessary to acknowledge that the Chinese system is built upon a long legacy of central planning and the CES, it is the willingness to set targets and hold the officials accountable that lies at the heart of the system's effectiveness in pursuing its objectives.

Table 4.6   *A summary of carbon governmental techniques and technologies at national level*

The technical	Juridical power	Disciplinary power	Government power
Planning and targets disaggregation	x	x	
Target responsibility systems	x	x	
Re-organization of the bureaucracy		x	
Re-ordered performance criteria and strengthened accountability	x	x	
Experimentation and piloting		x	x
Standards and quotas	x	x	x
Market-based mechanisms			x
Financial incentives and investment		x	x
Training, education and capacity building	x	x	x

5. Greening the financial system in China

INTRODUCTION

Making sufficient finance available is certainly one of the greatest challenges for mitigation and low-carbon transitions (MLCT) anywhere. In this regard, China's performance looks second to none. It invested a total of USD 758 billion, or 31 per cent of the global total, into renewable energy (RE) capacities between 2010 and the first half of 2019, at a 30 per cent compound annual growth rate (CAGR) between 2004 and 2018 (FS-UNEP 2019). Such a feat naturally raises a question: 'How has China managed to finance its MLCT at such a scale and speed?' Meanwhile, China has emerged as a global leader in green finance since the mid-2010s. It has reportedly introduced 'the world's most comprehensive set of national commitments, covering a range of priorities across banking, capital markets and insurance' in green finance (UNEP Inquiry/World Bank 2017, p. 24).

This chapter addresses the above question by exploring the relationship between China's MLCT investment and the government's efforts to green finance through the lens of the 'analytics of government' introduced in Chapter 2. My working hypothesis is that China has managed to finance its huge MLCT investment programme largely by greening its fiscal and financial systems through the development of a green finance governmentality.

This chapter has four main parts. The first part provides some contextual information about China's financial system. The second part explores the political rationalities of greening the financial system. It reviews the practices of various political authorities in creating the field of visibility, the forms of knowledge they drew on and generated, and the constitution of key subject groups and their transformation sought. The third section then examines the governmental techniques and technologies (GTTs) that have been deployed. The fourth section discusses the financing outcome. The chapter ends with concluding remarks.

CONTEXT: THE STATE'S STRATEGIC INTEREST AND ROLE IN THE FINANCIAL SYSTEM

China's efforts to green the financial system cannot be properly understood without recognizing the pivotal role of the state in the economy and the state's aim of accessing social capital while maintaining state dominance and control. Investment and accumulation have been key concerns in Chinese economic governance ever since the foundation of the PRC in 1949. It is deeply rooted in the CCP's Marxist roots. Brødsgaard and Rutten (2017) suggest that the imperative for the rapid accumulation of fixed capital derived from the Maoist strategy of industrialization and the Soviet conception of 'primitive socialist accumulation'. The latter holds that rapid industrialization under socialism needs to be sustained through state appropriation of surplus value from outside the state economy. While it was agriculture that was the target of such appropriation during the pre-reform era, since the corporatization of state-owned enterprises (SOEs) in the 1990s, it is the so-called 'social' capital that has been targeted. Through the corporatization of the SOEs and the strategy of 'grappling with the large and letting go the small', the government managed to turn around the financial fortunes of the SOEs and consolidate their leading positions in the economy (Zhang 2004). This has been aided by a process of financialization. Thus, financial institutions (FIs) have mushroomed. For instance, the number of banks had increased from just one (i.e., the People's Bank of China – PBC) in 1978 to three policy banks, five large commercial banks, 12 national joint stock banks, 133 city commercial banks, five private banks and 859 rural commercial banks by 2016 (Huang and Wang 2018). On the other hand, the capital markets have grown exponentially since the opening of the Shanghai Stock Exchange and Shenzhen Stock Exchange in 1991 and 1992, respectively. The market value of corporate and government bonds rose 151 times between 1997 and 2016 (from RMB 418 billion to RMB 63.7 trillion) (Huang and Wang 2018). By the end of 2018, domestic stock exchanges had a total of 3,584 listed companies, with a total market capitalization of RMB 43.5 trillion; the value of all outstanding bonds was RMB 85.98 trillion (PBC 2019). Globally, Chinese FIs and financial markets have become a force to be reckoned with. Chinese bond market capitalization was about 1 per cent of global GDP at the start of the 2000s, but became 9 per cent by the end of 2017; China's stock market capitalization rose from about 2 per cent of global GDP in 2002 to about 10 per cent in 2017. Most extraordinarily, the assets of the Chinese banking system 'skyrocketed from 1 per cent of global GDP in 2002 to about 40 per cent in 2017' (Cerutti and Obstfeld 2018, p. 6).

China's financialization is unique in that the government has acted increasingly as a shareholder, through its holdings in SOEs and state-owned banks and other FIs; it has simply added financial markets to its regulatory and planning toolkit (Wang 2015). The financial system has some unique characteristics. First, it is still led by banks, rather than capital markets. At the end of Q2 in 2019, Chinese FIs' total assets reached RMB 308.96 trillion, of which banking institutions' (excluding the central bank) total registered assets were RMB 281.58 trillion (or 91.1 per cent of the total registered assets) (PBC 2019). Second, it continues to be dominated by state-owned banks. As of 2012, the three policy banks and five state-owned commercial banks accounted for more than 53 per cent of total assets of the banking industry (PwC 2013). Third, the credit system is repressed. This refers to low real rates of return on deposits, which means that bank loans represent a subsidy from savers, mainly households, to those enterprises that can access such loans (Lardy 2008). Moreover, financial repression coexists with soft-budget constraint (SBC) for SOEs. Xu (2017) suggests that SBC lies at the roots of excess productive capacities and soaring leverage ratios by SOEs. Fourth and finally, the SOEs remain the main beneficiaries of this system, as they have privileged access to bank loans and capital markets. Indeed, this situation seems to have moved further in their favour after 2012: the share of SOEs in total bank credit to non-financial institutions (NFIs) increased from 35 per cent in 2013 to 83 per cent in 2016; between 2012 and 2018, assets of non-financial SOEs grew at more than 15 per cent annually (Lardy 2019).

GREENING THE FINANCIAL SYSTEM: POLITICAL RATIONALITIES

The efforts to green finance have evolved over a long period of time. Wang and Zadek (2017) identify three phases of green finance development in China: initiation (2007–10); consolidation (2011–14); and implementation (from 2015). However, the Industrial Bank Green Finance Group (IBGFG 2018) identifies four stages of this development after studying 59 central-level policy documents. These stages (with the number of documents issued in parentheses) are: germination (1981–2005) (10); initiation (2006–10) (21); take-off (2011–15) (17); deepening (2016 onwards) (10). The only substantive difference between the two classifications is that the IBGFG's analysis goes back further in time. Notably, measured by the number of documents issued by the central government, the period from 2006 and 2010 (i.e., the 11th FYP period under SOD influence) marked the most dense period of policy activity.

Creating the Field of Visibility

What and whom to be governed

The state has gradually expanded its field of visibility, spreading from banks and their lending activities to other kinds of FI and other forms of financial and informational instruments in the past four decades. Following the IBGFG's analysis, four distinctive phases can be identified. The first phase focuses on bank lending activities. The long 'gemination stage' (1981–2005) was influenced by the 1979 Law of Environmental Protection (trial), which was formally promulgated in 1989 (IBGFG 2018). Its key concern was to use the denial of bank credit as a weapon to protect the environment by curtailing the growth of 'low-level' industries. The 'Decision on Strengthening Environmental Protection during the Period of National Economic Readjustment' by the State Council (1981) required that large- and medium-sized construction projects that failed an Environmental Impact Assessment (EIA) should be denied bank credit. However, after 2003, under the influence of the SOD, the State Council's (2005) 'Decision on Implementing the Scientific Outlook on Development and Strengthening Environmental Protection' broadened the scope of intervention to include financial and fiscal policies as part of the coordinated mechanisms for positively protecting the environment. Working together with the central bank, the PBC, the Ministry of Finance (MoF) and the China Banking Regulatory Commission (CBRC), the State Environmental Protection Agency (SEPA) and its successor the Ministry of Environmental Protection (MEP) initiated discussions for an environmental tax and introduced a series of policy guidelines covering ecological compensation mechanisms, green trade, green government procurement, green insurance, green securities and green credits during 2007 and 2008 (Aizawa and Yang 2010). In other words, this second period was characterized by an attempt to use multiple financial instruments to meet the needs of environmental protection in the real economy (Zadek and Zhang 2014).

However, the institutionalization of Ecological Civilization after the 18th Party Congress in 2012, the renewed drive for reform in 2013, rising environmental awareness and the need for long-term MLCT investment in the wake of a growing environmental crisis motivated financial regulators to explore a more systemic approach during the 12th FYP (2011–15), ushering in a third phase. This period is characterized by the effort to transform the entire financial system to tilt resource allocation in favour of a resource-efficient and environment-friendly economy, as required by the notion of Ecological Civilization. The smog that affected large swaths of the eastern coastal regions in China in the winter of 2012 acted as a major impetus. It prompted the launch of an action plan by the State Council (2013) to prevent and control air pollution, setting a national cap on the consumption of coal as a proportion of total energy consumption (TEC), which was 'below 65 per cent by 2017'. The latter effectively boosted the demand for MLCT-related actions.

The desire to continue with economic reform has also promoted this drive. In November 2013, the 'Decision of the Central Committee of the Communist Party of China on Several Major Issues Concerning Comprehensively Deepening Reform' (CCPCC 2013) was announced. It argues that, to turn China into a strong, democratic, civil and harmonious modern country and to realize the 'Chinese dream' of national rejuvenation, it is necessary to deepen reforms and constantly enhance confidence in the road, theory and system of socialism with Chinese characteristics ('the three confidences' in short). It insists that development is the key to solving all problems and points out that the core of deepening reform is to place the relationship between the market and the government on the right footing. In this spirit, it calls for the establishment of 'a systematic and full-fledged institutional system of ecological civilization for the protection of the eco-environment' and 'a market-based mechanism that channels private capital investments to the protection of the eco-environment' (cited in GFTF 2015, p. 2). In this connection, establishing a green financial system is aligned with three major areas of reform of the financial system: developing capital markets; reducing resource misallocation by shifting from administrative supportive measures to reliance on market forces; and reducing financial risks by shifting from implicit to explicit government guarantees (Zhang et al. 2015).

On the other hand, two sets of international collaborative studies further encouraged the drive. In the post-2012 era, a range of policy institutions and financial regulators joined the discourse and knowledge creation around green finance. For example, between late 2013 and early 2015, the Finance Research Institute of the Development Research Centre (DRC) under the State Council formed an 18-month collaborative research project with the Canada-based International Institute for Sustainable Development (IISD) to explore policy options to 'support China in developing a "green financial system" and to encourage such developments internationally' (Zadek and Zhang 2014, p. 2). On the other hand, the Green Finance Task Force (GFTF), established in 2014, collaborated with the UNEP's Inquiry into the Design of a Sustainable Financial System (the 'UNEP Inquiry' hereafter). These two international initiatives produced three major reports and numerous background papers and helped inform decision-making.

Finally, the 'Guidelines for Establishing the Green Financial System' (hereafter 'the Guidelines') (PBC et al. 2016) and the 'Division of Work for the Implementation of the Guidelines to Establish a Green Financial System' (hereafter 'the Division of Work') (PBCRB 2018) heralded a fourth period, providing a clear definition of the field of visibility. The 'Guidelines' outline measures to promote the development of green finance covering bank lending, securities market, public–private partnerships (PPP), an environmental pollution liability insurance, environmental rights trading market and related financing instruments, local government initiatives and international cooperation. On the other hand, the 'Division of Work', which has not been made public in full, but is excerpted in the *China Green Finance Progress Report 2017* (PBCRB 2018), breaks down the programme of establishing a green financial system into 30 specific tasks, each with specified lead organization(s) and a deadline for implementation (p. 158). The detailed list, specifying 24 of the 30 tasks, illustrates an all-encompassing field of visibility, covering not only FIs (especially banks, insurance companies, and listed companies), financial markets covering equity, insurance, bonds, green credits, and other financial products, but also information sharing and other financial infrastructure, financial laws and regulations, and different forms of international collaboration.

The problems to be solved
The existing financial regime, as described above, has supported China's economic development over the past three decades and helped consolidate the dominance of state ownership in the key sectors of the economy. However, it has also had adverse consequences including rising environmental problems and the financial system's inability to provide sufficient long-term capital for environmental protection including MLCT. Zadek and Zhang (2014) identify six reasons for greening the financial system, including, among others, the financial markets' failure to invest in a sustainable economy, and, related to that, the systematic underestimation of financial risks for those assets that are associated with high-carbon activities. The GFTF (2015) links the environmental problems to a heavy industrial structure reliant on the use of coal and a pricing system that fails to reflect environmental externalities. It also points to declining growth rates of fiscal revenue and government expenditure, thus the need for private capital to meet between 85 and 90 per cent of China's green investment needs.

A heavy reliance on bank loans as a source of long-term investment was another problem. As of 2017, domestic bank loans funded 11.3 per cent of the fixed asset investment (FAI), compared with 6.1 per cent from the state budget, 0.3 per cent from foreign investment, 65.3 per cent from self-raised funds and 16.95 per cent from 'other sources' (NBS 2019). To make sense of this structure, it is necessary to point out that capital investment is subject to state regulation, which stipulates a minimum ratio of own capital for debt-financed FAI, thus effectively setting a limit on leverage ratios. For example, the 'Circular of the State Council on Adjusting the Capital Ratios for Fixed Asset Investment Projects' (State Council 2009) sets this minimum ratio among different sectors, ranging from 40 per cent for iron and steel to 20 per cent for housing and regular commercial residential projects. While this ratio offers some protection against financial risks, the credit-dominated system makes banks increasingly vulnerable to rising financial risks, especially in the context of climate change.

These problems are compounded by the fact that China is an investment-driven economy. Gross capital formation (GCF) as a percentage of GDP, GCF/GDP, has been increasing since the beginning of economic reform (or even earlier), although this trend seems to have attenuated from 2016. The GCF/GDP ratio has risen from 34.3 per cent in 2000 to 44.4 per cent in 2017, whereas the ratio of FAI to GDP has risen from 32.8 per cent in 2000 to 71.7 per cent in 2018 (after a peak of 81.9 per cent in 2016) (NBS 2019). However, the capital-output ratio (i.e., the amount of investment required to produce an additional output) has risen rapidly after a financial stimulus package was implemented in 2008 in the wake of the global financial crisis, reflecting declining efficiency of investment (Ma et al. 2018).

The objectives of greening the financial system

A range of objectives have been identified. The 'Guidelines' state that 'The main purpose of establishing the green financial system is to mobilize and incentivize more social (private) capital to invest in green industries, and to more effectively control investments in polluting projects' (p. 1). This highlights the overall objective of mobilizing capital and improving resource allocation according to environmental principles. The 'Guidelines' also refer to numerous other essential goals, including the sustainable development of the economy, establishment of a sound green financial system, improved functioning of the capital markets in allocating resources and aiding the real economy, and the construction of an Ecological Civilization. Moreover, it has been argued that the effort to green finance represents a step to make concrete intrinsic environmental values, as recognized by Chairman Xi's vision of 'clear waters and green mountains', and make them 'measurable, evaluable and tradable' (Liu and Zhang 2015).

A key objective has been to meet the long-term investment need for the transition towards a green economy including MLCT. Zheng (2015) provided detailed estimates from the NDRC for such investment for the period from 2015 to 2020. He considered the following seven categories: (1) pollution control (including urban environmental infrastructure and industrial pollution control); (2) electric power (including nuclear power, hydropower, solar power, power from other sources); (3) forestry; (4) waste recycling; (5) green build-ings; (6) electric vehicles; and (7) urban rail transit. His estimates were that the investment needs for the first five categories would grow from RMB 1,642 billion (USD 260 billion) (actual) in 2012 to RMB 2,867 billion in 2015 and RMB 2,908 billion in 2020, exceeding 3 per cent of GDP. Zheng's estimates show that energy conservation (EC) investment would claim the single largest proportion of this total investment (about 38 per cent in 2012 and 44.6 per cent in 2020). Zheng (2015) also estimated that, as of 2012, the sources of these green investments were 20 per cent self-owned capital, 10 per cent through public budgets and 70 per cent from market-based mechanisms. He pointed out that the 'slowdown of China's GDP growth rate and reforms to local government debt financing are putting pressure on the availability of fiscal funds for infrastructure development' (Zheng 2015, p. 54). Thus, he argued that it was necessary for 70 per cent of the future investment (i.e., about RMB 2 trillion a year) to continue to be financed by market-based green finance, in which he included loans, bonds, equities and other private financing.

Longer term large-scale investment needs have also been estimated. Citing the National Center for Climate Change Strategy and International Cooperation (NCCS), NDRC (2016) states that the estimated low-carbon investment requirement for the next 15 years is about RMB 30 trillion (or an annual average of RMB 2 trillion), including RMB 10 trillion for additional investment on energy conservation and RMB 20 trillion for additional investment on energy. The MEE (2018b) referred to a similar set of figures for 2016–30 also citing the NCCS as its source, stating that delivering China's 2030 mitigation goals and tasks requires an estimated cumulative investment of RMB 32 trillion (at 2015 constant prices) during 2016–30, or RMB 2.1 trillion every year on average. This total includes: RMB 13 trillion for additional energy conservation work, RMB 17.6 trillion for low-carbon energy development and RMB 0.13 trillion for developing forest carbon sinks. Thus, the latter estimation suggests more balanced investment needs between EC and RE development. On the other hand, the Research Centre for Climate and Energy Finance (RCCEF) of the Central University of Finance and Economics provided estimates of climate financing needs up to 2050 for meeting the key objective of China's Nationally Determined Contributions (NDCs) pledge under the Paris Agreement (RCCEF 2015). It suggested an annual investment need of approximately RMB 2.56 trillion (1.8 per cent of GDP) between 2020 and 2030. It anticipated that the need for climate financing would peak in 2020. A more recent study suggests that energy transition pathways consistent with the 2.0°C and 1.5°C climate goals will require new energy investment of RMB 3.3 trillion (1.5–2.0 per cent of GDP) and RMB 4.6 trillion (in excess of 2.5 per cent of GDP) per annum, respectively (He 2020).

On the other hand, it is possible to discern that China's efforts greening its financial system were partly motivated by the wish to enhance its international standing in sustainable development. As indicated in Chapter 3, China started to actively pursue international leadership from 2014. Pan Jiahua (Pan 2018, p. 529) observes: 'The year 2014 marked a strategic turning point in China's policy response to climate change. Proactivity replaced passivity, taking the lead replaced following others, domestic actions were ramped up and playing a leadership role on the international stage became an important objective.' In this context, greening finance seemed to offer China an opportunity to develop its international leadership in this growth area. For instance, the DRC-IISD project's interim report highlighted the leadership by emerging economies such as China and South Africa in greening the financial system (Zadek and Zhang 2014). This referred specifically to the CBRC's efforts in encouraging green credits. The project's final report again notes that 'China can play a leadership role in promoting sustainable development, especially by strengthening the international policies and regulation of green finance' (Zhang et al. 2015, p. 11). China went on to initiate, during its presidency in 2016, the G20 Green Finance Study Group (GFSG), which it co-chaired with the UK. This has led to significant policy learning by China (Zhang 2019).

Forms of Knowledge Drawn On and Given Rise To

Different forms of knowledge have informed China's governing effort to green its financial system. The first, it is cumulative knowledge about the extent of China's environmental problems. For example, researchers from the DRC under the State Council note that 'environmental damage poses significant risks to the country's quality of life, economic competitiveness, resilience and long-term growth. The estimated cost of pollution damage is between 3 per cent and 6 per cent of GDP and rising' (Zhang et al. 2015, p. 4). Similarly, the GFTF acknowledges that 'air quality is satisfactory in only 8 out of 74 major cities, and just 25 per cent of drinking water reaches national quality standards' (GFTF 2015, p. 2). Second, the economic notion of environmental externalities (in terms of both costs and benefits) has provided a shared vocabulary for various research efforts and policy discourses, whereas pre-existing research on the financial and human costs of the environmental problems has added weight to the effort to green finance (GFTF 2015).

Third, China learned from green finance discourse and policy practices from abroad. International cooperation has evidently facilitated learning about green finance. The DRC-IISD's interim report (Zadek and Zhang 2014) identifies an existing green finance approach among OECD countries that focuses on meeting the environmental protection needs of the real economy and addressing short-term biases, misaligned incentives and better stewardship of assets, but is unable to generate enough long-term investment for MLCT. In contrast, the authors argued strongly for an alternative, systemic approach towards the greening of the financial system. The project's final report sketches out a green financial system that defines its goal as being aligned with sustainability and encompasses three key elements: a policy system that internalizes environmental externalities; FIs that respond to the signals represented by such externalities; and financial products that serve as instruments for managing risk and intermediating capital (Zhuo and Zhang 2015, p. 30). Similarly, the GFTF's (2015) final report also argues for a systemic approach towards the greening of the entire financial system. Both the DRC-IISD research project and the GFTF involved numerous well-placed international partners (Elliott and Zhang 2019).

Moreover, Chinese researchers and policy makers have demonstrated the ability to learn, that is, the capacity to assimilate information and adjust the means and goals of their own projects accordingly. For example, the joint DRC-IISD team endeavoured to make a definitional distinction between narrowly defined and broadly defined green finance initiatives. They pointed out that, while the former 'seek to identify what proportion of a particular set of financial assets or institutions can be considered green' and is popular among OCED countries, the latter 'seek to define the overall goal of the financial system in terms of sustainability and provide a means to assess its effectiveness. They set criteria in terms of the purpose of the financial system in allocating capital efficiently and effectively in light of environmental risks. The focus of broad definitions is the stability of the whole financial system and macro-economy' (Zhang et al. 2015, p. 18). The Chinese authors advocated for the adoption of the broad definitions. Moreover, on this basis, they justified considering all investments that improve the efficiency of fossil fuel use and reduce energy consumption intensity as 'green' (Zhang et al. 2015, p. 19). Meanwhile, Chinese researchers sought to highlight the intrinsic link between green finance and Chairman Xi's vision of 'clear waters and green mountains' (Zhuo and Zhang 2015).

The work of the GFTF is another example of action-oriented and externally leveraged learning. According to Jun Ma (GFTF 2015), a co-convenor of the GFTF, the taskforce was established at his proposal to capitalize on the discussions and exchanges that took place at the Eco Forum Global in July 2014 in Guiyang, China. Ma was the chief economist of the PBC's Research Bureau at the time. However, his former career included eight years with the IMF and World Bank and 14 years with Deutsche Bank. Ma's account shows that the process of establishing the GFTF was organic, as the membership 'expanded from 20 or so at the start to more than 40 individuals' over a six-month period, during which the group put together a set of recommendations. In the end, 14 recommendations were selected out of 19 topics that the GFTF explored and presented in the final report. These covered four areas: specialized investment institutions; fiscal and financial policy support; financial infrastructure; and legal infrastructure. The report by the GFTF, published in April 2015, echoes the DRC-IISD team's call for greening the financial system. It argues that a green finance system 'would allow China to attract private capital into green industry, reduce fiscal pressure, create a new growth area and enhance economic growth, stability and restructuring' (GFTF 2015, p. 2). The report further explains that this system would also provide economic incentives to spur green investment and curb brown investment by increasing the returns on investment of green projects, reducing the returns on investment of polluting projects, and enhancing investor, business and consumer awareness and responsiveness to these signals (GFTF 2015, p. 2).

Significantly, such learning became institutionalized through the 'Integrated Reform Plan for Promoting Ecological Progress' (hereafter 'the Integrated Plan') by the CCP Central Committee and the State Council in September 2015. The Integrated Plan calls for the establishment of a green finance system to support the pursuit of ecological progress. In one paragraph, it includes all but one of the 14 recommendations by the GFTF (2015). The only recommendation that was not taken up by the government was about establishing a green bank on the model of the UK's Green Investment Bank. Rather the decision was to incorporate the principle of green finance into the work of all FIs (Wang 2018). Moreover, the 'Guidelines' (for greening the financial system) announced in August 2016 by seven top Chinese ministries/commissions (PBC et al. 2016) distinguishes green finance from a green financial system. While green finance refers to financial services provided for economic activities that are supportive of environment improvement, climate change mitigation and more efficient resource utilization, the 'Guidelines' state, a green financial system is 'the institutional arrangement that utilizes financial instruments such as green credit, green bonds, green stock indices and related products, green development funds, green insurance, and carbon finance, as well as relevant policy incentives to support the green transformation of the economy' (PBC et al. 2016, p. 1). From this point onward, China has entered a new phase where the focus is firmly placed on a comprehensive transformation of the financial system.

With its active stance on green finance, China has become a significant contributor to new knowledge about green finance, further enriching its practice. For example, led by the Green Finance Committee (GFC), the Green Finance Book Series has published 13 titles, covering green finance policy, Chinese and international case studies of green finance development, green credit, green bonds, green funds, environmental risk analysis, green technologies, green assessment, environmental, social and governance (ESG) investment, the 'belt and road' green investment and the like (Ma 2020). Moreover, in collaboration with the GFC of the China Society for Finance and Banking, the Industrial and Commercial Bank of China (ICBC) (2017) developed stress testing methods to explore the impact of environmental factors on credit risk as well as bringing the idea of an ESG rating index and Green Index to wide international attention. The supervision workstream of the Network of Central Banks and Supervisors for Greening the Financial System (NGFS), led by Ma Jun, published an 'Overview of Environmental Risk Analysis by

Financial Institutions' and 'Case Studies of Environmental Risk Analysis Methodologies' in September 2020.

Constitution of Subjects and Their Transformation

The greening of the financial system has broadened the field of visibility from FIs and their investment approaches and activities to include capital markets such as the inter-bank bond market and the stock exchanges, insurance companies, publicly listed companies and other financial service providers. China made its first attempt at introducing the principle of green credit back in 1995 in the form of a 'Circular on Implementing Credit Policy and Strengthening the Work on Environmental Protection' (PBC 1995). The Circular stressed that FIs should take account of borrowers' performance in environmental protection and pollution treatment in their loan-making decisions. It required that FAI projects that fail the EIA be denied credit. It recommended differentiated lending policy towards four types of projects and enterprises. These include the 'prohibited' and the 'restricted' according to the state's industrial policy, those promoting environmental protection or benefiting the environment, and those engaging in environmental protection and pollution treatment. However, it 'did not come to implementation', as it was a time when the overriding policy emphasis was still on economic development (Aizawa and Yang 2010).

Such effort was stepped up under the influence of the SOD (State Council 2005) and thanks to the inclusion of two binding targets for *jieneng jianpai* in the 11th FYP (2006–10) (see Chapter 3). The 'Opinions on Implementing Environmental Protection Policies and Rules to Prevent Credit Risks' (hereafter the 'Opinions'), which were jointly issued by the SEPA, PBC and CBRC (SEPA et al. 2007), sought to reactivate the earlier effort. These advocated differential treatment for projects classified by the state industrial policy as 'encouraged', 'restricted' and 'eliminated'. Moreover, the 'Opinions' called for the establishment of an information communication and sharing mechanism between environmental protection agencies (EPAs) and financial authorities. While the EPAs were required to share information on enterprises' environmental records with local branches of the PBC, CBRC and FIs through regular bulletins, FIs were instructed to strengthen their credit management by using this information when reviewing loan applications. The SEPA's (2008) 'Circular on Standardizing the Provision of Enterprise Environmental Violation Information to the Credit Reporting System of the People's Bank of China' sought to standardize the sharing of such information by FIs on credits extended to those enterprises. As shown below, a series of follow-up measures have been introduced to make banks not only more responsible, but also more capable and incentivized to green their lending activities. On the other hand, stock exchanges and the inter-bank bond market have been mobilized to support the issuance of green bonds, starting in 2016 (Zhang 2019). Finally, publicly listed companies have been put on a three-year programme to institutionalize their environmental information disclosure.

GREENING FINANCE: GOVERNMENTAL TECHNIQUES AND TECHNOLOGIES

A series of GTTs have been developed and adapted to encourage and incentivize green financial products and behaviours. Some of these take rather traditional forms. For example, a study of 59 green finance related central-level policy documents identifies 23 circulars, 4 decisions, 4 management methods, 3 action plans, 12 sets of guiding opinions, 2 comprehensive plans, 1 set of institutional building measures, 3 guides, and so on (IBGFG 2018). It reveals that while circulars and decisions dominated in earlier stages, action plans and overall plans were more prevalent in later stages. However, other measures are radical and innovative. Below, we highlight six key GTTs in the wider programme of greening finance, starting with the greening of the fiscal system.

Fiscal Support

Public finance has increased its support for energy conservation and environmental protection (ECEP), and under it, renewable energy (RE) and energy conservation (EC). The budgetary category of 'environmental protection' was renamed as 'energy conservation and environment protection' in 2010. The principal techniques here are RE and EC allocations within the General Public Budget (GPB) and RE subsidies paid out of the Renewable Energy Development Special Fund (REDSF) within the General Funds Budget (GFB), mainly funded by the RE Levy revenue (see Chapter 4 for background). Data on detailed budget lines have become available only from 2008. Moreover, ECEP covers a dozen sub-areas. Hence it is difficult to pin down the exact public resource allocations for MLCT. Nevertheless, the information presented in Table 5.1, which includes data from both the China Statistical Yearbook and annual budgetary accounts on the MoF's website, is illuminating. It shows that ECEP spending has grown from RMB 128.17 billion in 2008 to RMB 739.02 billion in 2019 – at a 17.3 per cent CAGR, compared with a 13.1 per cent CAGR for overall GPB spending. Meanwhile, ECEP's proportion in overall GPB has risen from 2.09 per cent to 3.09 per cent during the same period. These figures testify to the prioritization of ECEP in the GPB.

Furthermore, the table shows significant budgetary flows to RE and EC during 2008–19, totalling a cumulative amount of RMB 133.4 billion and RMB 637.8 billion respectively. While these show a strong focus on EC within the GPB, combining the budgetary allocation with the spending funded by the

RE Levy brings the total public spending on RE to RMB 539.21 billion, comparable with the cumulative spending of RMB 524.93 billion on EC, during 2012–19. This means that the two streams attracted similar amounts of input from public finance over 2012–19. The table also shows contrasting trends on budgetary spending on RE and EC. During 2008–19, the budgetary allocation to EC increased on average by 17.9 per cent CAGR, whereas the same rate was minus 0.7 per cent CAGR for RE. However, RE Levy-funded expenditure grew by 28.8 per cent CGAR over 2012–19. Finally, budgetary allocation for RE and EC peaked respectively in 2013 and 2015. Another trend is that, from 2014 onwards, the amount of funds allocated to these two streams have shown large fluctuations. Meanwhile, growth of the RE Levy-funded expenditure has also slowed – it grew by only 2.4 per cent in 2019. Comparing the average annual spending during the 12th FYP with the same for the first four years of the 13th FYP shows that the growth rates are respectively 70.2 per cent for EPEC, minus 56.9 per cent for RE, 6.3 per cent for EC and 106.3 per cent for RE Levy-funded expenditure.[1] All these show that the support for MLCT investment by public finance is not being sustained and increasing reliance is placed on the RE Levy, whose growth is also slowing. This explains to some extent why the mobilization of private (social) capital became so crucial from the mid-2010s.

---

[1]   The last percentage represents changes from 2012–15 to 2016–19. See Table 5.1.

Table 5.1   MLCT-related public expenditure from General Public Budget and RE Levy, 2007–2019

Year	GPB (RMB bn)	GPB YoYG (%)	ECEP (RMB bn)	ECEP YoYG (%)	RE (RMB bn)	RE YoYG (%)	EC (RMB bn)	EC YoYG (%)	RE Levy (RMB bn)	RE Levy YoYG (%)
2007	4978.135		99.582 .							
2008	6138.6	23.3	128.17	28.7	8.5		11.08			
2009	7629.993	24.3	193.404	50.9	5.901	-30.6	19.698	77.8		
2010	8987.416	17.8	244.198	26.3	11.788	99.8	40.193	104.0		
2011	10924.779	21.6	264.098	8.2	14.16	20.1	43.944	9.3		
2012	12595.297	15.3	296.346	12.2	14.755	4.2	53.874	22.6	14.615	
2013	14021.210	11.3	343.515	15.9	19.706	33.6	68.204	26.6	28.231	93.2
2014	15178.556	8.3	381.564	11.1	14.662	-25.6	58.065	-14.9	44.843	58.8
2015	17587.777	15.9	480.289	25.9	16.471	12.3	83.346	43.5	57.96	29.3
2016	18775.521	6.8	473.482	-1.4	8.612	-47.7	62.265	-25.3	59.506	2.7
2017	20308.549	8.2	561.733	18.7	5.299	-38.5	66.828	7.3	71.209	19.7
2018	22090.600	8.8	629.761	12.1	5.673	7.1	64.564	-3.4	83.879	17.8
2019	23885.837	8.1	739.02	17.3	7.895	39.2	67.779	5.0	85.895	2.4
Total (RMB bn) (2008–19)	178124.135		4735.58		133.422		639.84			
Total (RMB bn) (2012–19)	144443.35		3905.71		93.073		524.925		446.14	
CAGR (2008–19) (%)	13.1		17.3		-0.7		17.9			
CAGR (2012–19) (%)	9.6		13.9		-8.5		3.3		28.8	
Total (RMB bn) (2011–15)	70307.62		1765.81		79.75		307.43		145.65 (2012–15)	

Year	GPB (RMB bn)	GPB YoYG (%)	ECEP (RMB bn)	ECEP YoYG (%)	RE (RMB bn)	RE YoYG (%)	EC (RMB bn)	EC YoYG (%)	RE Levy (RMB bn)	RE Levy YoYG (%)
Total (RMB bn) (2016–19)	85060.51		2404.00		27.48		261.44		300.49	
Change in annual spending from 2011–15 to 2016–19 (%)	51.2		70.2		-56.9		6.3		106.3*	

Note: YoYG = year-on-year growth; ECEP = Energy Conservation and Environment Protection; RE = renewable energy; EC = energy conservation.
*This shows change from 2012–15 to 2016–19. RE Levy denotes expenditure funded by the RE Levy.
Sources: http://www.stats.gov.cn and http://yss.mof.gov.cn.

Green Credit Policy, Green Credit Statistics and Green Bank Evaluation

Following the joint recommendation by the SEPA, CBRC and the PBC (SEPA et al. 2007) to implement environmental protection policies and rules, the CBRC (2007) issued 'Guidance on Lending for Energy Conservation and Emissions Reduction' (hereafter 'the Guidance'). This was the first policy document issued by Chinese financial regulators that bears the expression of 'energy conservation and emissions reduction' (*jieneng jianpai*) in its title. It is subsequently referred to as China's original 'green credit policy' (Aizawa and Yang 2010; Yu et al. 2013). The Guidance stresses the importance of *jieneng jianpai* and the role of bank loans in supporting *jieneng jianpai*. It highlights the need to prevent financial risks associated with high energy consumption, high pollution and technologically backward economic activities, which became known as 'two highs, one surplus' industries and activities. It reinforces the message of differentiated lending treatment for different types of activities according to the state's industrial policy (see Chapter 4 for details) and as specified by the joint recommendations (SEPA et al. 2007). Other stipulations of the 'Guidance' are:

1. All loans must meet six necessary conditions for lending, including an EIA and energy conservation audit (ECA) (a new requirement).
2. Loans are to be managed by categories (A, B, C), depending on their environmental impact.
3. Borrowers with major energy consumption and pollution risks shall be put on a separate entity list.
4. Differentiated interest rates shall apply to *jieneng jianpai* projects.
5. *Jieneng jianpai* lending shall become an important criterion in the evaluation of FIs. It shall be linked to the performance evaluation of FI senior managers and have a bearing on administrative decisions on the FI's market access and business development plans.

Furthermore, the 'Guidance' requires FIs to treat the promotion of *jieneng jianpai* as an important mission and a concrete manifestation of their corporate social responsibility. It further specifies steps to be taken by banks in areas of strategic planning, internal audit, risk management and business development. It also offers guidance regarding credit policy and management, outlining a series of concrete action points. Finally, the 'Green Credit Guidelines' (GCG) issued by CBRC (2012) took another step forward. They decree annual reporting of green credits to the Board of Directors and the CBRC, disclosure of information covering green credit strategy, policy and development, bi-annual comprehensive review of green credit performance and the submission of a self-assessment report to the CBRC. The guidelines also outline a series of requirement on institutional capacity building and the development of processes and procedures. This was followed by the introduction of a nation-wide Green Credit Statistical System (GCSS) in July 2013 (CBRC 2013).

The pressure on banks to green their practices was intensified by the introduction of the Green Credit Business Self-Assessment (GCBSA) for the 21 major banks administered by the CBRC (2014). The GCBSA capitalizes upon the GCSS and is based on an evaluative system called 'Key Performance Indicators of Green Credit Implementation' (KPIGCI). The latter comprise 142 key indicators, with 77 'qualitative' and 65 'quantitative'. The quantitative indicators are in turn divided into 39 core indicators and 26 optional indicators, measuring the quality of loans in terms of *jieneng jianpai* under nine headings. In contrast, the qualitative indicators cover areas such as organization and management, institutional capacity building, procedural management, internal control and information disclosure, supervision, and examination. As part of the GCBSA, the 'two highs, one surplus' industries were specified to cover 29 sectors. As dictated by the 'Division of Work', the China's Banking Association (CBA) was subsequently delegated by the CBRC to organize and implement the evaluation of green banks. On 26 December 2017, the CBA issued the 'Implementation Plan of Green Bank Evaluation in China Banking Industry (Trial)' and a 'Green Bank Evaluation Scoring Sheet' (see Table 5.2). The latter follows the same categories and indicators as those in the KPIGCI (CBRC 2014) and attaches individual weightings to the indicators.

Moreover, in July 2018, the PBC (2018) launched the Green Credit Performance Evaluation of Banks and Deposit-Taking Financial Institutions ('GCPE' in short) for 24 major banks, further reinforcing the system's calculability and comparability. The GCPE metrics comprise two sets of indicators – the quantitative and qualitative – and assign respectively a weighting of 80 per cent and 20 per cent to them. The quantitative set consists of five indicators: (internal) proportion of green loan balance; share of green loan balance in the industry; ratio of green loan balance growth to total loan balance growth; year-on-year growth of green loan balance; ratio of bad loans. The qualitative set consists of three indicators: implementation of national green development policy, implementation of the GCSS, and implementation of the CBRC's GCBSA. This design enables the GCPE to build on the CBRC's work on GCSS and GCBSA.

Table 5.2   *Green Bank Evaluation Scoring Sheet*

Category of indicators	Level-1 indicators	Weighting (%)	Level-2 indicators	Evaluation items*	Score points
Qualitative	Organizational management	30	Board of Directors' responsibilities	2.7.1–2.7.3	12
			Senior management responsibilities	2.8.1–2.8.6	10
			Dedicated department	2.9.1–2.9.2	8
	Institutional capacity building	25	Policy making	3.10.1–3.10.4	8
			Management by category	3.11.1–3.11.5	5
			Green innovation	3.12.1–3.12.5	5
			Own performance	3.13.1–3.13.6	2
			Capacity building	3.14.1–3.14.6	5
	Procedural management	25	Due diligence investigation	4.15.1–4.15.7	5
			Compliance appraisal	4.16.1–4.16.5	3
			Credit approval	4.17.1–4.17.4	3
			Contract management	4.18.1–4.18.2	5
			Fund payment management	4.19.1–4.19.3	3
			Post-loan management	4.20.1–4.20.4	5
			Off-shore project management	4.21.1–4.21.5	1
	Internal control and information disclosure	15	Internal examination	5.22.1–5.22.4	5
			Appraisal and evaluation	5.23.1–5.23.3	5
			Information disclosure	5.24.1–5.24.4	5
	Supervision and examination	5	Self-assessment	6.26.1–6.26.2	5

Category of indicators	Level-1 indicators	Weighting (%)	Level-2 indicators	Score points
		Total		100
Quantitative	Change in the sum of loans in two areas**	–	3 points for positive year-to-year growth / No points for negative year-to-year growth	A max of 3 additional points
	Completion of quantitative KPIs and optional indicators	–	2 points for completing all sections / 1 additional point for completing all KPIs, but incomplete on optional indicators / Zero points for incomplete information on both KPIs and optional indicators	A max of 2 additional points

Note: *Corresponding to those in CBRC (2014). **These two areas are: (1) energy conservation and environmental protection projects and services; (2) energy conservation and environmental protection, alternative energy, and alternative energy vehicles.
Source: CBA (2018, No. 171).

Mandatory Reporting and Information Disclosure by FIs

Taking advantage of its central planning legacy, China has established a comprehensive informational and performance monitoring infrastructure regarding green finance, covering: regular reporting, statistics compilation, information disclosure, performance self-assessment and external evaluation. The reporting system through the GCSS and the KPIGCI played a pivotal role until 2017. It required major banks and local banking regulatory authorities to compile and submit statistics every six months. The 21 major banks must report their green credit statistics directly to the CBRC, with January to June of 2013 as the first reporting period (CBRC 2018a). In January 2018, the PBC stepped in to further strengthen the reporting system by issuing the 'Circular on the Establishment of a Green Credit Special Statistical System' (GCSSS, in short) (PBC 2018), effective from March 2018. Although this document has not been made public, it is understood from the PBC's annual report (PBC 2017) that the GCSSS seeks to be consistent with the GCSS administered by the CBRC since 2013. However, it covers three more banks (Bank of Beijing, Bank of Shanghai and Bank of Jiangsu). It also requires more frequent reporting of the statistics, at quarterly intervals. Moreover, it serves as the basis for the GCPE.

Information disclosure is being rolled out across the financial industry. Consistent with the GFTF's recommendation and the 'Division of Work', the latest focus is on publicly listed companies. This builds upon the 'Environmental Information Disclosure of Listed Companies (Trial)' by the MEP (2010). According to Ma Jun, this specified three steps designed to establish a system of compulsory environmental information disclosure by listed companies: (1) major emitting listed companies were required to disclose their environmental information by 2017; (2) all listed companies were required to disclose such information a on semi-voluntary basis by 2018; (3) compulsory disclosure was to be imposed on all listed companies (including FIs) by 2020 (GFC 2018). This was corroborated by Fang Xinghai, Deputy Chair of the China Securities Regulatory Commission (CSRC). Fang confirmed in March 2017 that the CSRC issued a circular in December 2016, stating that, from 2017 onwards, those listed companies that belong to key emitting sectors named by the MEP must disclose their emissions in the previous year in their annual reports (Zhu 2017). In December 2017 the CSRC issued 'Guidelines for the Disclosure Contents and Formats of Listed Companies'.

However, information disclosure on the use of green bond proceeds and their environmental impact is inadequate. CBI and CCDC (2019) found that while post-issuance reporting on fund allocation is available from 94 per cent of Chinese green bond issuances, the level of detail in the information on environmental impact disclosure was wanting. Yu and Li (2017) reported that, while three-quarters of the 80 disclosure reports they examined meet the regulatory requirements, of the three key areas of disclosure – use of proceeds, project progress and realized environmental effects, the extent of disclosure was lowest for the last one. Two causes were suggested. First, the four regulators (PBC, CSRC, NDRC, and the National Association of Financial Market Institutional Investors (NAFMII)) had different levels of requirement for such disclosure. Second, the specific forms and key reporting points were different (Yu and Li 2017). Another factor is the dominance in green bonds issuance by large banks, who tend to only provide information on the allocation of their green loans between different categories of sectoral activity, rather than project-level environmental impacts. Escalante et al. (2020, p. 4) found that only 114 out of the 233 issuances prior to September 2018 that they examined had publicly available reporting of environmental impacts.

On the other hand, such disclosure appears to have relatively limited effects on lenders. Wang et al. (2019) used panel data from 320 companies in heavy polluting industries listed on the Shanghai Stock Exchange from 2008 to 2016 to study the relationship between level of information disclosure and corporate green financing. They found no significant positive correlation between the two and regarded this as evidence that there is collusion between the enterprises and local governments, through the latter's possible influence on banks' lending decisions. An alternative interpretation could be that the quality of the information is too poor to be useful for influencing credit decision-making by FIs.

Setting Green Finance Standards and Taxonomies

China was relatively slow in defining what is 'green'. The GCG, China's ground-breaking green credit policy document, did not specify green credit. It was not until the announcement of the GCSS (CBRC 2013) that the definition of green credits started to emerge (see Table 5.3). Moreover, a unified set of standards is lacking. The PBC and the NDRC have set their own standards on green bonds, covering FIs and SOEs, respectively. Labelled green bonds (LGB) in China were kicked off on the 22 December 2015, when the PBC (2015) made an announcement on 'Matters Concerning the Issue of Green Financial Bonds in the Inter-Bank Bond Market'. A 'Green Bond Endorsed Project Catalogue' (GBEPC), developed by the GFC, was annexed to the announcement. The GBEPC identifies six categories of eligible project (see Table 5.3). Subsequently, the NDRC (2015) and CSRC (2017) also issued guidelines, for SOEs and publicly listed corporations, respectively. While the CSRC (2017) followed the standards set by PBC, the NDRC (2015) identified 12 categories of eligible project in its Green Bond Issuance Guidelines (GBIG) (Table 5.3). Having started with defining discrete green financial products and projects, China then moved on to define green industries. In March 2019, seven key ministries jointly issued the 'Green Industry Guidance Catalogue' (GIGC) (NDRC et al. 2019). It identifies six level-one categories, 30 level-two sub-categories and 211 level-three sub-sub-categories (see Table 5.3). The GIGC differs from earlier guidelines and catalogues issued by the NDRC, PBC and CBRC in that it covers entire industrial chains, rather than targeting individual projects (Lu et al. 2019). It aligns closely with the state's industrial policy and the objective of 'lightening' the industrial structure through green finance. This fragmented regulative framework has caused some confusion and practical difficulties.

Another problem is the misalignment between China's green bond standards and various international standards. There are two issues here. On the one hand, unlike most international standards, both the GBEPC and GBIG consider clean coal and efficient utilization of fossil fuels to be valid uses of green bond proceeds. On the other hand, the GBIG allows LGB issuers to use up to 50 per cent of the proceeds for refinancing and topping up operational capital, whereas international organizations such as the Climate Bonds Initiative (CBI) define green bonds as those bonds with at least 95 per cent proceeds dedicated to green assets and projects (CBI 2019a). The two stock exchanges in China have also adopted relaxed green bond standards, allowing up to 30 per cent of the proceeds to be used this way (SZSE 2018; Xinhua Finance 2019; Escalante et al. 2020). In the first half of 2019, only 49 per cent of the LGBs issued in China met international green bond standards (CBI 2019b). These have caused significant disquiet among international stakeholders (e.g., Pronina 2019). Fortunately, while Chinese researchers sometimes attempted to defend China's position on this in the past, the consultation edition of the updated GBEPC (2020) proposes to remove clean utilization of fossil fuels to maintain GBEPC's 'advanced nature, to better connect with relevant international standards, and to raise China's discursive power on green bond standards' (PBC et al. 2020).

Table 5.3   *Taxonomies of green credit, green bond, and green industry*

Green finance entity	Existing standard	Project/industry standard scope
Green credit	GCSS (CBRC 2013, 2014)	13 categories: green agriculture; green forestry; energy and water conservation and environmental protection industrial projects; nature protection, ecological repair and disaster prevention and mitigation; resource circular use; waste treatment and pollution prevention; renewable energy and clean energy; rural and urban water projects; construction energy conservation and green building; clean transport and transit; energy conservation and environmental protection services; off-shore projects that adopt international conventions or standards; strategic and new industries (energy conservation and environmental protection; new energy; new energy-driven vehicles)
Green bond	GBEPC (PBC 2015)	Six categories (and 31 sub-categories): energy conservation, prevention and treatment of pollution; resource conservation and recycling; clean transport; clean energy; ecological protection; climate adaptation
	GBIG (NDRC 2015)	12 categories: energy conservation; green urbanization; fossil fuel energy cleaning and efficiency improvement; new energy; circular economy; water resource conservation and non-conventional water resource exploitation; pollution prevention and treatment; ecology and forestry; energy conservation and environmental protection industrial projects; low-carbon industrial projects; eco-civilization pilots and demonstration and experimentation projects; low-carbon development pilots and demonstration projects
Green industry	GIGC (NDRC et al. 2019)	Six categories (and 30 sub-categories): energy conservation and environmental protection industries, clean production industries; clean energy industries; eco-environment industries; greening and upgrading of infrastructure; green services.

Source: Adapted from Lu et al. (2019).

Administrative and Financial Incentives for Green Bonds and Green Credits

What epitomized the development of green finance in China in the 13th FYP period is an explosive growth of the green bonds market from 2016 (Zhang 2019). The easing of administrative control on market access to green bonds, the inclusion of green bonds as qualified collaterals for medium-term lending facility (MLF), and the incorporation of green finance performance into the Macro Prudential Assessment (MPA) are key measures.

The government's incentive for green bonds started with selectively relaxing administrative control on their issuance. The principal offer of the three key policy documents on green bonds issued by the PBC (2015), NDRC (2015) and CSRC (2017) was that applications for green bond issuance would enjoy speedy approval procedures (*jibao jipi* in Chinese), which effectively shortened the application process from three months to one month (Zhang 2019). The rules issued by the two stock exchanges (Shanghai and Shenzhen) regarding the issuance and trading of pilot green bonds in the spring of 2017 were in similar spirit. In comparison with the gradualist approach to the promotion of green credits, the government's approach to green bonds was more like a 'big bang'. Shortly after the announcement of the PBC's (2015) guidelines on the 22 December 2015, four banks applied for approval to issue China's first ever LGBs. By the 25 January 2016, just a month later, two of these banks – the Industrial Bank (IB) and Shanghai Pudong Development Bank (SPDB) – were each given permission to issue LGBs worth up to RMB 50 billion (EID 2016).

Furthermore, the GBIG issued by NDRC (2015) and applied to SOEs include several features that constitute a relaxation of the rules set out in the 'Opinions on Strengthening Risk Prevention of Enterprise Bonds' (NDRC 2014b). These include:

1. The ratio of debts/total investment can be up to 80 per cent (rather than 70 per cent).
2. Issuers are not constrained by quotas of debt issuance.
3. When the debt/asset ratio is lower than 75 per cent, the size of an issue is not affected by the size of the issuer's other debts.
4. If secure repayment measures are in place, issuers can use up to 50 per cent of the proceeds for repaying bank loans and topping up operating capital.

Moreover, the incorporation of LGBs and green credits into the MPA has provided monetary incentives to the banks for either issuing green bonds or making green loans. The green bond profile of the 24 major banks was incorporated into the MPA in 2016; from the third quarter of 2017, the green credit performance of the 24 major banks also became incorporated into the MPA (PBC 2017, p. 22). The pre-existing MPA framework is a quarterly evaluation system of banks (Zheng 2018). In its current form (effective from January 2017), it consists of 18 indicators under seven equally weighted categories, with each indicator carrying its unique weighting. The framework sets three performance classes for the banks: A for 'excellent' with a score of 90 or more; B for 'qualified' with a score of 60 or more but below 90; C for 'unqualified' with a score below 60. It also features an incentive mechanism. A bank with an 'A' score can gain an extra 10 per cent revenue of its reserve deposits' income at the PBC, whereas a bank with a 'C' score will lose 10 per cent of its reserve deposit interest income. Since the required reserve ratio in China is high (16.5 per cent for large banks and 13.5 per cent for small and medium banks), this mechanism represents a strong incentive. Moreover, the outcomes of the MPA may also affect the banks' other businesses, in particular innovation applications and approvals (Zheng 2018). Finally, according to Ma Jun (2018), by March 2018, the PBC had also incorporated green bonds as qualified collateral for MLFs, set up a standing loan facility, pledged supplementary lending and refinancing, and was looking to do the same for green credits in terms of MLFs and refinancing in accordance with the 'Division of Work'.

A dozen SNGs have also taken action to encourage green bonds. These measures fall into two main categories: (1) positive policy signals including issuing special documents, establishing a taskforce, and launching pilot programmes; and (2) financial incentives including cost sharing, credit enhancement and investment guidance. For example, the Futian District government in Shenzhen decided in 2017 to provide a 2 per cent subsidy for corporate green bonds (SGF and CBI 2017). The Zhongguanchun Administration in Beijing decided in April 2017 to subsidize 40 per cent of coupon investment payment for corporate green bonds (up to RMB 1 million per year) (SGF 2017).

International Cooperation

China started its collaboration with the outside world on green finance back in the mid-2000s, prompted by the adoption of the SOD. Its first ever green credit product was a loan fund for *jieneng jianpai* projects, offered jointly by the IB and the International Finance Corporation (IFC) in 2006 (Zhang et al. 2015, p. 6). The IFC offered to share (50:50) with the IB any potential losses on such loans. Moreover, it was during this process that the IB learnt about the Equator Principle (EP) and discovered green finance as a growth area.[2] China developed its original 'Green Credit Policy' in collaboration with the EP in

---

[2]   Interview 19 September 2018.

2006 (Aizawa and Yang 2010). Subsequently, the IB became the first Chinese bank to adopt the EP in 2008. The Bank of Jiangsu and the Bank of Huzhou became EP banks in 2017 and 2019, respectively. Besides, as shown earlier, the international collaborations between the IISD and DRC, and between the GFTF and the UNEP's Inquiry, directly informed the 'Guidelines' (PBC et al. 2016) that set a national framework for greening the financial system.

By 2015, China's rationality of greening its finance was bolstered by its 'go global' strategy. China's presidency of the G20 Summit in September 2016 significantly stimulated China's drive to green finance through international engagement. In the run-up to the summit, China proposed to establish the Green Finance Study Group (GFSG) under the G20, which was adopted by the G20 Finance and Central Bank Deputies meeting in December 2015. The GFSG was co-chaired by China and the United Kingdom, represented respectively by the PBC and the Bank of England. It was supported by the UN Environment as its secretariat. Ma Jun was the co-chair on the Chinese side. The 'Guidelines' were issued on the 31 August 2016, days before the G20 summit in Hangzhou. The activities of the GFSG continued under the Germany presidency of G20 in 2017 and the Argentina presidency in 2018, although the GFSG was renamed the 'Sustainable Finance Study Group' (SFSG) in 2018, as the USA wanted.[3] After two years' inactivity, in March 2021, the SFSG was revived under the Italian presidency, with China and the USA co-chairing.

China has become an emerging global leader in green finance. It is one of the eight founding members that established the NGFS in December 2017 (NGFS 2017). Initially established to circumvent friction with USA in the G20 GFSG (and SFSG),[4] the NGFS (2019) has expanded its global coverage to include more than 60 countries. Ma Jun from China leads one of the three work streams, on 'supervision'. China is also a member of the International Platform on Sustainable Finance (IPSF), launched by the European Commission in October 2019. Aiming to develop a 'common group taxonomy' for its member jurisdictions, the IPSF has initiated a working group on taxonomies, starting with those of the EU and China (CBI 2021). Moreover, through the Working Group of UK–China Climate and Environmental Information Disclosure Pilot (WGUKCCEIDP), China also plays a pivotal role in implementing and further developing the recommendations of the international Task Force on Climate-Related Financial Disclosure (TCFD) through piloting (TCFD 2017). By the end of 2019, this scheme involved 14 participating FIs, includ-ing the ICBC, IB, Bank of Jiangsu, Bank of Huzhou, Efunds, China Asset Management, PICC, Ping An, and AVIC Trust on China's side, and HSBC,

---

[3]   Interview 18 September 2018.
[4]   Interview 18 September 2018.

Aviva, Hermes, Brunel Pension Fund, and Environmental Agency Pension Fund on the UK side; the Huzhou National Green Finance Reform Leading Group, representing one of China's five green finance pilots (see Table 4.5), acts as an observer (UKCCEIDP 2020). Finally, China has also taken an active role in the development of the international framework standard on sustainable finance (ISO/TC322), with Ma Jun serving as vice chair of the working group. The Research Centre for Green Finance Development of Tsinghua University and the Energy Conservation and Low Carbon Group in China National Institute of Standardization are involved.[5] Through these steps, China has found a valuable niche in green/sustainable finance for its pursuit of international leadership.

OUTCOMES

Let us now examine the effectiveness of this green finance governmentality. We will do so by focusing on the contributions to MLCT investment. Determining the size of the investment into the LCT in China is not a straightforward task due to the diversity of sources (RCCEF 2017). Here a preliminary assessment is made to explore the correspondence between governing efforts and the contributions to MLCT investment by different kinds of FIs and instruments. Following Zheng (2015), we assume that, during 2012–19, China invested 3 per cent of its GDP into MLCT. This amounts to an estimated total of RMB 17,857.6 billion (approximately USD 2,744.6 billion) or RMB 2,232 billion (approximately USD 343 billion) per annum. Several observations can then be made.

First, China depended mainly on domestic sources, although overseas issuance of green bonds has raised a significant amount of capital. During the 12th FYP period, the contribution of the Clean Development Mechanism (CDM) was quite limited: in its peak year (2013), CDM generated USD 1.212 billion; its share in China's total RE investment in 2015 (USD 115.4 billion) was only 0.2 per cent (Li et al. 2017). In comparison, during 2016–19, overseas issuance of green bonds raised a total of RMB 172.4 billion (approximately USD 25.6 billion), equivalent to 15.8 per cent of all the capital raised from green bonds at the same time (Chen and Jiang 2020).[6]

Second, investment into EC more than doubled between the 11th FYP and the 12th FYP periods, but grew only moderately from the 12th to the 13th FYP

---

[5] See https://www.bsigroup.com/en-GB/our-services/events/webinars/2020/isos -sustainable-finance-standards/.
[6] Climate Bonds Initiative (CBI) and China Central Depository & Clearing Research Centre (CCDC Research) (2020) data put these numbers higher: USD 33.8 bn or RMB 240 bn in total, equivalent to 19.7 per cent.

period. According to Li et al (2017), during the 12th FPY period (2011–15), energy efficiency improvement (EEI) attracted a total investment of USD 324.8 billion (approximately RMB 2,036.5 billion) (2.7 times that of the 11th FYP), including USD 49.4 billion (approximately RMB 309.7 billion) (15.2 per cent) from public finance (2.2 times that of the 11th FYP).[7] In comparison, during the 11th FYP period (2006–10), public investment into EEI was USD 22.45 billion (18.7 per cent), leveraging in USD 97.97 billion of private investment. Our calculation in Table 5.1 shows that the average annual investment from public finance for EEI increased from the 12th FYP period to the first four years of the 13th FYP period by only 6.3 per cent.

Third, investment into RE grew rapidly in the 12th FYP period, but declined in the 13th FYP period. Over the periods 2011–15 and 2016–19, China invested a total of USD 376 billion (approximately RMB 2350 billion) and USD 429 billion (approximately RMB 2885 billion), respectively, into RE according to FS-UNEP (2019, 2020).[8] The average growth rates for the two periods were respectively 22.9 per cent and minus 4.7 per cent GAGR. Coupled with Table 5.1, it can be estimated that public finance, including both direct budgetary allocation for RE and from the RE Levy revenue, provided respectively RMB 225 billion and RMB 328 billion for RE investment during the two periods, contributing 9.6 per cent and 11.4 per cent, respectively. They grew by 39.4 per cent and 8.3 per cent GAGR respectively. Evidently, the REDSF became the principal source of public spending on RE during the 13th FYP period. China's RE investment peaked in 2017 at USD 143 billion (RMB 994 billion). The investment in 2019 was USD 83.4 billion, down by nearly 42 per cent from the peak level (FS-UNEP 2020).

Fourth, the dependency on bank loans is evidently high, but different for the two streams, EC and RE. Li et al. (2017) point out that, during the 12th FYP period, leveraged by FiT-related subsidies, RE investment relied on asset-backed mortgages, although the extent of this reliance varied between different forms of renewable energy. For example, for wind energy projects, bank loans accounted for 80 per cent of total investment, with the rest from own capital. In comparison, the investment in energy efficiency improvement relied more on subsidies from public finance. The sources for EEI in 2015 were: 17.5 per cent from government, 31.1 per cent by self-raised funds, 43.9 per cent from bank loans, 5.9 per cent from equities and 1.8 per cent from other sources (Li et al. 2017, p. 30). Clearly, bank loans are the most impor-

---

[7]   The Chinese currency amounts are estimated by applying an average yearly exchange rate over the five-year period at USD 1 = RMB 6.27.
[8]   This is estimated by applying the yearly exchange rates.

tant source of finance for both EC and RE despite the difference in degree of dependency.

Fifth, the contribution of the domestic carbon market has been very limited. In 2015, its cumulative trading value was only RMB 750 million (RCCEF 2017). Nevertheless, the total transaction volume rose from 160 million tonnes at the end of 2016 to 256 million tonnes at the end of 2018, whereas the value of such transactions arose from RMB 2.5 billion (IIGF and RCCEF 2017) to RMB 13.989 billion during the same period (Goepe 2018). The contribution of traditional international financial markets is not large either: in 2016, RMB 3.5 billion was raised through IPO abroad, while China invested a total of RMB 10.3 billion in renewable energy globally (RCCEF 2017).

Sixth, green bonds have raised a significant amount of capital for MLCT investments, mainly during the 13th FYP period. Research shows that, between 2016 and 2019, more than a trillion RMB has been raised by this method (Chen and Jiang 2020). However, green bonds make up less than 1 per cent of the total bond market in China (Escalante et al. 2020). Moreover, this development has mainly benefited FIs and SOEs. Among the onshore issuances, which are the mainstay, the FIs and SOEs claimed respectively 65 per cent and 32 per cent of the amount raised, compared with 3 per cent for the private sector (Escalante et al. 2020).

Seventh and finally, by far the most important source for financing the MLCT has been green credits. Moreover, they have grown steadily. Citing a CBRC source, Zhang et al. (2011) reported that the total amount of loans granted by the five largest commercial banks in support of *jieneng jianpai* was only RMB 10.6 billion in 2007. By 2011–12, green loans by the 16 largest commercial banks amounted to approximately RMB 1.7 trillion a year.[9] ICBC, China's largest bank, alone made green loans worth RMB 593 billion at the end of 2012 (Yu et al. 2013). The growth of green credit has accelerated from 2013, when the CBRC introduced the GCSS and defined green credits. In March 2018, the CBRC published the multi-year statistics (see Table 5.4). It shows that from June 2013 to June 2017, the value of the green credit balance of the 21 major banks rose from RMB 4.85 trillion to RMB 8.30 trillion, a growth of 14.3 per cent CAGR. Of these, the green credit balance for 'energy conservation and environmental protection' (Type 1) and 'strategic emerging industries' (Type 2) increased from RMB 3.69 trillion to RMB 6.53 trillion (17.5 per cent CAGR) and from RMB 1.51 trillion to RMB 1.69 trillion (5.5 per cent CAGR), respectively (CBRC 2018a). Between December 2013 and June 2017, green credits to renewable energy and clean energy grew from

---

[9]   This is estimated according to Yu et al. (2013) by calculating the average of the two years' numbers.

RMB 1.0407 trillion to RMB 1.6103 trillion, causing a cumulative energy saving of 1.2 billion tce and reducing CO_2 emissions by 3.1 GtC (China Bond Rating Ltd 2018). Green credits have continued to rise since the launch of the GCSSS. The end-of-year green credit balance has risen from RMB 7.1 trillion in 2018 to RMB 11.95 trillion in 2020 (PBC 2019; FBIL 2021), at 13.9 per cent CGAR. Year-on-year growth from 2019 to 2020 was at 16.9 per cent. However, it appears that the proportion of green credits in total loans has not increased substantially from 2013 to 2020. Table 5.4 shows that the ratio of green credits by the 21 banks to total bank credits was 6.78 per cent, 6.93 per cent, 7.05 per cent, and 6.70 per cent, respectively, for 2013, 2014, 2015, and 2016. On the other hand, the ratio of green credit balance by the 24 major banks to total loan balance has reportedly risen from 6.45 per cent in 2019 Q1 to 6.88 per cent in 2020 Q4 (FBIL 2021).

To summarize, the contributions of four principal sources to the estimated MLCT investment spending in China during 2012–19 are as follows: budget-ary allocation at RMB 618 billion (3.5 per cent); RE Levy at RMB 446 billion (2.5 per cent); green bonds at RMB 1,000 billion (6.6 per cent); and green credits at RMB 9,000 billion (50.4 per cent). Together these sources accounted for an estimated 62 per cent of all the funding, with the majority of the rest (38 per cent) likely to be funded by investors' own capital. Interviews suggest that bank loans for technological upgrading are additional to green credits, thus representing an additional source of finance. This means that the governmental efforts on green credit and LGBs have had the biggest impact on financing China's MLCT. It is also evident that, to a large extent, China has been able to finance its MLCT in the past decade not by substantially raising the weighting of green credits, but by ensuring that a significant proportion of bank credits, at least 6–7 per cent, continues going to MLCT-related activities.

Research into the behavioural changes by Chinese FIs is still in a state of infancy. Nevertheless, by November of 2017, the membership of the GFC (formed in April 2015) came to manage two-thirds of all China's financial assets (Wang and Zadek 2017). As far as the banks are concerned, they have responded unevenly. In a study of 12 banks in 2008, Zhang et al. (2011) found that while all the banks emphasized their commitment to denying loans to 'two-highs, one surplus' industries as required by the CBRC, only two banks had introduced detailed implementation policies and mechanisms. Progress remains highly uneven. Comparing the data compiled by Yu et al. (2013) and those from the Global Green Finance Big-Data Platform, it is possible to compile a league table of green credits by China's top banks. By the end of 2018, the top seven banks were (with outstanding green credit balance in parentheses): ICBC (RMB 1,237.758 billion), Agricultural Bank of China (RMB 1,050.4 billion), China Construction Bank (RMB 1,042.26 billion), IB (RMB 844.9 billion), Bank of China (RMB 632.667 billion), Bank of

Table 5.4 Green credit statistics for 21 major banks (2013–2017) (RMB 100 m)

Indicator	Jun 2013	Dec 2013	Jun 2014	Dec 2014	Jun 2015	Dec 2015	Jun 2016	Dec 2016	Jun 2017
Type 1 credits	34293.74	36853.49	41610.42	44363.86	49734.66	53201.57	55728.24	58090.34	65312.63
Type 2 credits	14233.1	15129.6	15606.85	15764.44	16626.67	16864.56	16907.05	16956.52	17644
Total green credits (Types 1&2) (A)	48526.84	51983.09	57217.26	60128.29	66361.33	70066.13	72635.29	75046.87	82956.63
6-month growth (%)		7.12	10.07	5.09	10.37	5.58	3.67	3.32	10.54
YoYG (%)			17.91	15.67	15.98	16.73	9.45	7.11	14.21
Memo items									
All loans (B)		766327		867866		993460		1120552	
YoYG (%)		13.89		13.25		14.47		12.79	
Ratio (%) (A/B)		6.78		6.93		7.05		6.70	

Note: All loans (B) here refer to outstanding loans provided by all banking institutions. YoYG = year-on-year growth.
Sources: CBRC (2018a) for green credits by the 21 major banks; CBRC (2018b) for all loans.

Communications (RMB 283 billion), and SPDB (RMB 217.515 billion). However, their respective growth rates, at 109 per cent, 590 per cent, 335 per cent, 650 per cent, 178 per cent, 129 per cent (2011–18), and 753 per cent, tell a quite different story. These differences show that some of the largest banks have not performed as well as their public profile would suggest. For example, ICBC's green credit balance increased from RMB 914.603 billion in 2015, to RMB 978.56 billion in 2016 and RMB 1,077.199 billion in 2017. However, the proportion of the bank's green credit in its total credit balance hardly changed, standing at 7.66 per cent, 7.49 per cent and 7.72 per cent in 2015, 2016, and 2017 (WGUKCCEIDP 2018). In contrast, the EP banks have performed better. By the end of February 2018, the IB had provided green credits for 14,621 enterprises, with a cumulative value of RMB 1.4836 trillion; its amount of outstanding green credits was nearly RMB 700 billion, accounting for 20 per cent of all its loans to enterprises (IBGFG 2018, p. 88). The Bank of Jiangsu's green credit balance grew from RMB 24.3 billion in 2015 to RMB 46.74 billion in 2016 and RMB 66.97 billion in 2017, with their ratio in all comparable loans rising from 5.20 per cent to 9.10 per cent and 12.20 per cent (WGUKCCEIDP 2018).

CONCLUSION

Analysis in this chapter shows that China's fast-paced and large-scale investment in MLCT has been accompanied by the development of a green finance governmentality. In terms of political rationalities, the field of visibility covers all types of FIs and financial instruments, although banks, stock exchanges, the inter-bank bond market, and listed companies have received the most attention. It also includes public finance, the RE Levy, and the REDSF. The main objectives include: increasing investment in green industries while reducing investment in 'two highs, one surplus' industries, mitigating climate-related financial risk for FIs, and finally exploring opportunities to develop China's international leadership in green finance. The governmentality draws on a variety of knowledge, including most importantly the notion and extent of environmental externalities, green finance principles and practices abroad, and those arisen from China's own experimentation with green finance since the mid-2000s.

It is also evident that this governmentality has followed closely the wider rationalities of the SOD and Ecological Civilization. It started with a moderate ambition of protecting the environment and reducing the associated financial risk in the mid-1990s with limited effects. However, the combination of the SOD and the inclusion of mandatory *jieneng jianpai* performance targets in the 11th FYP strengthened this rationality by emphasizing coordinated development, making green finance an indispensable element of implementing the SOD in general and the state's industrial policy in particular. Moreover, from 2015–16 onwards, the programme of greening the financial system has become coupled with the institutionalization of Ecological Civilization and the 'go global' rationality, gathering further support both at home and abroad.

China's effort to green its financial system has been significantly aided by the development and deployment of numerous GTTs. The latter include increased allocation from public finance, the creation of a dedicated REDSF and a FiT scheme supported by the mandatory RE Levy, a green credit policy consisting of regular and frequent reporting, self-assessment and external evaluation of banks based on regularly compiled statistics, administrative and monetary incentives for green bonds and green credits, green finance standards and taxonomies, mandatory information disclosure by FIs, and international collaboration. Evidence shows that despite the gaps in the implementation of standards, taxonomies and information disclosure, the system has effectively contributed to China's MLCT investment. Green credits, LGBs and the REDSF, the targets of much governmental effort, have financed more than half of all the estimated investments in MLCT. Green credits also supported the growth of SEIs including the EV industry.

To a large extent, the analysis in this chapter validates my hypothesis that the greening of the financial system, underpinned by its own governmentality, has played a crucial role in China's success in financing its MLCT.

6. Localizing the low-carbon transition: a tale of three Chinese cities

INTRODUCTION

I know when and where the seeds of this book were sown. The date was 23 March 2013 and the place was Qingdao Liuting International Airport. I was on my first visit to this northern Chinese city, made famous by its namesake beer (Tsingtao Beer).[1] I had just stepped over from the airbridge into the terminal building. Almost right in front of me, there was an advertisement board several metres wide and about two metres high. It featured two men, both in suits, one Chinese, looking like an official, and the other, a Westerner, with a friendly smile. They were apparently in conversation. The Chinese man eagerly asked: 'Is it possible to reduce a city's emissions by 30%?' The Westerner answered 'Yes, naturally'. It caught my eye because it was the first close sight I got of the place. Moreover, the caption was in English and the subject coincided with the purpose of my visit, which was to study low-carbon development in the city. In the ensuing years, the memory of that advertisement has stayed fresh. For me, it indicates one thing above all else: city officials really matter in climate mitigation in China.

Given China's highly decentralized economic and administrative system, it is impossible to fully understand China's performance in MLCT without looking at what happens at the sub-national level. The system of 'setting targets centrally and implementing targets locally' for energy conservation and emissions reduction (ECER or *jieneng jianpai* in Chinese) has made Chinese cities the key carriers of the nationally and provincially disaggregated mitigation targets (WRI 2011a, p. 30). They have become the 'executive location from which to create a low-carbon economy' (Wang et al. 2015, p. 82). This chapter thus explores the experiences of three Chinese cities: Qingdao of Shandong province, Shanghai, and Hangzhou of Zhejiang province. Where relevant, it also looks into the experiences of the two related provinces (Zhejiang and Shandong). Shandong is the largest emitter of CO_2 of all Chinese provinces; in terms of carbon intensity, Shandong lies in the middle range, whereas Zhejiang

---

[1]   Qingdao is spelled Tsingtao in its former official romanization.

and Shanghai are among the lowest in China (ZCCCLCDC 2015). The focus of my investigation is how authorities in these cities have addressed the MLCT challenge by simultaneously responding to central policies and meeting local needs through the process of 'translation', which is about forging alignment between central directives and local projects (Rose 1999). I will explore four key questions in this chapter. First, what political rationalities do the cities have in pursing MLCT? Second, what governing techniques and technologies (GTTs) do they use to promote MLCT? Third, how have they financed MLCT? Fourth and finally, how have they performed so far?

This chapter is organized as follows. The first section introduces the case cities. The second section reviews the rationalities that the cities have adopted, including the creation of the field of visibility, including defined problems and objectives, knowledge creation and utilization, and the targeted social groups and their transformation. The third section provides insights into the GTTs that the relevant authorities have mobilized, either in response to central directives or acting on their own initiatives. The fourth section investigates the question of finance and investment. The fifth section examines the outcomes. The final section concludes.

INTRODUCING THE THREE CASES

The three cases have several features in common. First, they are all situated in China's most developed eastern coastal region and belong to the league of so-called 'first-tier' cities in socio-economic development terms on the Chinese mainland, with a per capita GDP around USD 20,000 (see Table 6.1 for summary statistics). Second, their economies remain reliant on industry, as the secondary sector still contributes between 30 per cent and 40 per cent of their GDP. Third, they are front runners in China's MLCT effort. All three cities are part of the national Low-Carbon City Pilots (LCCP) programme (cf. Chapter 4). Fourth and finally, they have all pledged to peak their emissions by 2020, ten years before the national target year (GLCDTTP 2016).

On the other hand, there are also significant differences between these cities. First, they have different administrative statuses. While Shanghai is a province-level municipality, the other two cities are at vice-province level. Hangzhou is the capital city of Zhejiang province, whereas Qingdao is one of five 'separately listed' cities in the centralized planning system. A study of these municipalities thus enables us to discern potentially what happens at the governing interfaces between the centre and provinces, between cities and their provinces, and between the cities and their urban district authorities. Another inter-city difference lies in population size and density. While both Hangzhou and Qingdao have a population of less than 10 million, Shanghai has a population of 24.3 million (2018). On the other hand, population density in Shanghai is about five times higher than those in the other two cities. Third, in terms of economic structure, the three cities are situated at different stages of de-industrialization, with Shanghai most advanced and Qingdao least so. Fourth, their economic growth rates also vary. Although they all grew quickly by international standards, ranging from just below 7.5 per cent to 9.7 per cent per annum in terms of GDP during 2011–15, Qingdao and Hangzhou has grown significantly faster than Shanghai has over the past decade. Fifth and finally, their fiscal resources and capacities also differ, reflective of both the size of their respective economies and administrative status. In 2018, Shanghai's General Public Budget (GPB) revenues was as high as RMB 711 billion (21.8 per cent of its GDP), compared with RMB 123.19 billion (9.1 per cent) and RMB 182.51 billion (15.2 per cent) for Qingdao and Hangzhou, respectively. Nevertheless, Qingdao and Hangzhou's overall revenues were as high as RMB 370.55 billion (QMBS 2019) and RMB 345.75 billion (HMBS 2019), respectively.[2]

Table 6.1   *Socio-economic indicators of Shanghai, Qingdao, and Hangzhou*

Indicators	Shanghai	Qingdao	Hangzhou
Area size (km²) (2018)	6,340	11,282	16,596
Permanent resident population (million) (2018)	24.23	9.39	9.81
Population density (persons/km2) (2018)	3,822	832	591
Permanent resident population growth (2010–18) (%)	5.2	23.0	42.6
GDP composition by industry (%) (2018)	0.6:29.8:69.9	3.6:40.0:56.4	2.3:33.8:63.9
GDP per capita (USD) (2018)	20,393	19,405	21,184
Average growth rate of GDP p.a. (RMB) (2006–10) (%)	11.1	13.8	12.4
Average growth rate of GDP p.a. (2011–15) (%)	7.5	9.7	9.1
Average growth rate of GDP p.a. (2016–18) (%)	6.8	7.6	8.1
General Public Budget revenue (RMB billion) (2018)	710.82	123.19	182.51

Note: GDP per capita figures in USD (2018) for Qingdao and Shanghai are converted from RMB amounts at the official average exchange rate (USD 1 = RMB 6.62); p.a., per annum.
Sources: SMBS (2011, 2016, 2019); QMBS (2019); QMPG (2011, 2016a); HMBS (2016, 2017, 2019); HMPG (2016); ADB (2018).

The three cities also have rather different energy structures, although all three are energy importers, as is characteristic of Chinese coastal cities. As of 2012, Qingdao's most important source of primary energy was oil (47.6 per cent), rather than coal as it was for Shanghai and Hangzhou (GLCDTTP 2016). This

---

[2]   Similar information is not available for Shanghai.

represents a further increase from 39.42 per cent in 2008 (WRI 2011a, p. 43). This is because oil and oil products are used as raw material inputs into the petrochemical production in Qingdao (WRI 2011b, p. 129). Another commonality is that they are all short of renewable energy (RE) resources. As of 2010, even if all such resources were utilized, RE would account for less than 9 per cent of Qingdao's energy consumption (WRI 2011b, p. 93). Hangzhou's energy self-sufficiency was only 3 per cent (Cheng and Guo 2016, p. 208). As of 2013, Hangzhou had the highest level of dependency on electricity imports at 26.6 per cent; coal contributed 36.8 per cent of its final energy consumption (GLCDTTP 2016). In comparison, Shanghai had the heaviest reliance on coal, at 41.4 per cent of total primary energy consumption.

MUNICIPAL POLITICAL RATIONALITIES

A review of the policy documents from the three cities over the past decade shows that the municipal authorities have been influenced by a variety of political rationalities. But the administrative rationality apparently dominates over other rationalities. They have tried to align their existing development strategies with the new low-carbon imperative and have started to build a knowledge base on emissions, standards, and options of mitigation and investment. Their target groups are similar, including mainly officials and MEUUs, although Hangzhou has paid attention to individuals and households from the outset.

Let us take Shanghai first. In the 13th FYP for Energy Conservation and Climate Change, Shanghai Municipal People's Government (SMPG) (2017a) reflected on its own shortcomings at four levels and called for stronger action on *jieneng jianpai*. First, internationally, it acknowledged that its global city comparators such as New York, London and Tokyo had set more ambitious mitigation goals. Second, nationally, it highlighted that the central government had called for Ecological Progress, green development and the peaking of emissions by 2030. Third, internally, it noted the city's decision to control overall population size and to reduce the amount of land used for buildings due to space constraint. Fourth and finally, it referred to the expectations and wishes of the city's residents for liveability and higher-quality urban life. This shows that Shanghai is influenced by multiple rationalities ranging from the global (competitiveness) and administrative, to the territorial and biopolitical. By 2019, echoing what is in the 'Integrated Reform Plan for Promoting Ecological Progress' (CCPCC and State Council 2015a), the Shanghai Municipal Leading Group on Climate Change and Energy Conservation and Emissions Reduction (SMLGCCECER 2019) was speaking of a re-orientation towards 'defining heroes by output per unit of land, efficiency, energy efficiency, and the quality of the environment', instead of GDP.

Hangzhou has shown a strong interest in the concept of a low-carbon city, the vision of 'Beautiful China' and the developmental opportunities that MLCT could bring. As early as 2009 and before the city became a LCCP, the city authorities put forward a comprehensive set of 'six-in-one' low carbon development goals, seeking to integrate low-carbon development into the economy, buildings, transport, lifestyle, environment and society (HCCPMC and HMPG 2009). Explaining why it is necessary to address climate change, Hangzhou's Climate Change Plan (2013–20) (HDRC 2014) provides three principal reasons. First, the state and province have raised higher requirements on addressing climate change. Second, there is increased environmental constraint and climate change-related pressure on the socio-economic development of the city. Third, there exist major developmental opportunities in climate change-related work including funding for potential investment, and the scope to strengthen the local economy and leverage for wider development. The plan proposes a strategy of distinguishing Hangzhou by its environment (*huan jing li shi* in Chinese) and developing a 'Hangzhou model of Beautiful China'.

Finally, Qingdao has tried to explore synergies between the green and low-carbon economy and the so-called 'blue economy' that it has been pursuing since 2010. The vision of 'blue economy' aims to develop Shandong Peninsula into an economic powerhouse of national significance by exploiting its coastal location and ocean-based resources; it will also make Qingdao the leader of this regional economy according to a plan approved by the NDRC (NDRC 2011c). The city government acknowledges: 'Low-carbon development has great significance for the strategic goal of pioneering scientific development, realising the blue transition and accelerating the construction of a liveable, happy, modern international city' (QDRC 2014, p. 10).

Meanwhile, the political and administrative logics are apparently dominant. Thanks to the incorporation of the binding MLCT indicators in the FYPs, the three cities are compelled to meet the targets assigned to them by higher authorities. Both Shanghai and Qingdao explicitly committed themselves to 'meeting nationally disaggregated and allocated targets' for reducing energy intensity and CO_2 emissions intensity in their 13th FYP (SMPG 2016a; QMPG 2016a). Moreover, there is a ratchet effect. In the case of Hangzhou, Zhejiang province was required by the centre to reduce carbon intensity by 19 per cent during the 12th FYP period, compared with the national target of 17 per cent. In turn, Zhejiang province required Hangzhou to reduce carbon intensity by 20 per cent during the period, and to achieve this target a year ahead (HDRC 2014). Hangzhou actually cut its carbon intensity by 21.9 per cent within four years (2011–14) (see Table 6.3).

The cities have started to build up a knowledge base on emissions and mitigation options. Of the three cities, I managed to obtain a detailed GHGs inventory only for Qingdao, although Hangzhou should have the most comprehensive and up-to-date inventories thanks to its status as a national pilot in GHG inventory compilation (see Chapter 4). Qingdao's first full GHG inventory was compiled by a group of experts in collaboration with the municipal authorities during 2011–12 as part of a technical assistance project funded by the Asian Development Bank (ADB) and coordinated by the World Resources Institute (WRI). The inventory shows that the composition of energy-related CO_2 emissions, without considering electricity imports, in 2009 was: industry (30.65 per cent), transport (21.7 per cent), power (19.57 per cent), buildings (16.19 per cent), heat (10.62 per cent), and agriculture (1.26 per cent) (WRI 2012). If indirect emissions from heat and electricity are taken into account, the respective contributions were industry (50.1 per cent), buildings (26.1 per cent), transportation (22.2 per cent), and agriculture (1.6 per cent). The contributions of the growth of population, GDP per capita, energy intensity and carbon intensity of energy to the growth of total CO_2 emissions over 2005–09 were found to be respectively 5 per cent, 63 per cent, minus 28 per cent and 4 per cent. These figures suggest that industry and economic growth were the most important contributory factors to emissions growth (WRI 2012, p. 48).

Furthermore, the report explores two scenarios: (1) keeping constant 2010 CO_2 emissions intensity for the next 10 years; (2) achieving a 45 per cent reduction of CO_2 emissions intensity from 2005 to 2020, and an 18 per cent reduction from 2010 to 2015. It finds that five major energy intensive industries – power and heat, chemicals, iron and steel, petrochemicals, and nonmetal production – consumed 63.7 per cent of total industry-related energy but contributed only 14.3 per cent of sectoral value added in 2009 (WRI 2012, p. 29). Most importantly, by taking into account the mitigation scope and cost-effectiveness of different options, the report recommends the following key areas of action for the second scenario (their expected contribution to mitigation in parentheses): restructuring the three industries by increasing the contribution of the tertiary industry to 60 per cent (24.29 per cent); adjusting industrial and product structures (16.68 per cent); upgrading industrial technology and improving energy efficiency measures (21.79 per cent); developing low-carbon, green buildings (10.06 per cent); constructing low-carbon urban transport systems (6.28 per cent); optimizing energy mix (18.76 per cent) (including raising the share of electricity imports (9.56 per cent)) and developing RE (9.20 per cent) (WRI 2012, pp. 59–60).[3] The report further identifies a total of 44 specific technologies and estimates their marginal average cost in 2020 under the 45 per cent target scenario (WRI 2012, p. 117). It also finds that, under the 45 per cent

---

[3] The percentages in parentheses show the proportion of emissions reduction that could be achieved by each action.

emissions intensity reduction scenario, by 2020, transport would become the largest source of emissions in Qingdao, at 29.3 per cent, compared with 28.4 per cent for buildings and 19.0 per cent for industry (WRI 2012). This study established a crucial baseline for Qingdao and developed a technical roadmap for Qingdao for the period of 2010–20. As we will see later, the results of the study directly informed the city's Low-Carbon Development Plan (LCDP), published in 2014.

In Shanghai, although no strategic LCT plans exist, civil society organizations and academics based in Shanghai's numerous universities have led the development of a knowledge base. As early as 2011, a research group led by Professor Dai Xing-Hai of Fudan University and Professor Zhu Da-Jian of Tongji University explored a low-carbon roadmap for Shanghai up to 2050 (WWF-SLCDRRT 2011). Another group of researchers led by Professor Tao Xiao-Ma of Tongji Univerity conducted a roadmap study for developing a low-carbon economy in Shanghai under the sponsorship of the Shanghai Association of Science and Technology (Tao et al. 2013). Wu et al. (2016) from Fudan University have estimated the levels of hypothetical carbon taxation required to peak GHGs before 2030 in Shanghai under two scenarios: (1) 'new normal' growth; (2) high growth associated with accelerated development of the Shanghai Free Trade Zone (FTZ) under three different carbon tax-level scenarios (low, medium and high). The low carbon-tax scenario refers to a base carbon price of RMB 30/t-CO_2, which was the average carbon price on the CET pilots in China in 2015, growing at 5 per cent per annum between 2016 and 2030. The medium carbon-tax scenario refers to a base rate of RMB 40/t-CO_2, growing at 10 per cent per annum from 2016 to 2030. The high carbon-tax scenario refers to a base rate of RMB 50/t-CO_2, growing at 15 per cent per year from 2016 and 2030. Under the high scenario, the carbon tax level would reach RMB 354/t-CO_2 in 2030. In other words, carbon prices would have to rise by 10 times between 2016 to 2030. Their studies suggest that, to peak its CO_2 emissions by 2020 while fully exploring the economic growth potential associated with the FTZ, it would be necessary for Shanghai to adopt the high carbon-tax scenario. This illustrates how much carbon prices would have to rise if exclusive reliance is placed on them.

It is evident that local officials and researchers in the three cities have shown a strong awareness of the significance of the MLCT process and have taken steps to create and accumulate knowledge about emissions and the potential of different policy tools. This becomes clearer when their GTTs are examined.

MUNICIPAL GOVERNMENTAL TECHNIQUES AND TECHNOLOGIES

Looking across the three cities, ten types of carbon GTT can be identified. Let us examine them below.

Building New Institutional Capacity

As Chapter 4 shows, the entire administrative structure from the State Council at the centre to provinces was adjusted around 2007/2008 to introduce a three-pronged administrative structure for tackling climate-related issues in China: an interdepartmental Leading Group (LG) on energy conservation (EC), pollution reduction and climate change, led by the top administrative head; a dedicated department responsible for related policy making and implementation; and a secretariat of the LG in the national and provincial Development and Reform Commission (DRC).

My fieldwork in the three cities reveals that substantial institutional capacity building has taken place at both the city and urban district levels. In Qingdao, the municipal Energy Conservation Office (ECO) was established within the municipal DRC during 2009–10, with an established staff (*bianzhi* in Chinese) of 11 employees.[4] This followed the introduction of the mandatory EC target in the 11th FYP in accordance with of the Energy Conservation Law (ECL) (WRI 2011b). Furthermore, within the Government Offices Administration (GOA) of Qingdao, which oversees general affairs in the public sector, an ECO was also set up in May 2009. The GOA established its own LG on EC in public institutions, with its own secretariat and a Section of Energy Conservation Management. This section had a staff of three people in 2013. Its responsibilities included decision-making, planning, supervision and assessment, and guidance regarding EC work in the 2,000-plus public sector institutions in the city. The budget for this three-person section increased from RMB 200,000 in 2010 to RMB 300,000 in 2013.[5] In accordance with the 12th FYP, an annual plan on EC was issued and distributed to each public sector institution, although the public sector accounted for only 6–7 per cent of all emissions in the city. The point was to demonstrate the public sector's leadership in cutting emissions in line with central policy.[6] Given that CO_2 emissions from public buildings more than doubled from 2005 to 2009 in Qingdao (WRI 2012, p. 47), this emphasis on public institutions made good sense. Another related depart-

---

[4] Interview 24 March 2013.
[5] Interview 28 March 2013.
[6] Interview 28 March 2013.

ment within the Qingdao municipal DRC is the Qingdao Energy Conservation Supervision Centre. Its establishment is dated to September 2010 on its website. It has a staff of 42 employees in eight sections (Baidu 2021a). Its functions include: supervision of EC, energy audit, review of energy use status report, energy impact assessment (EIA) for fixed asset investment (FAI) projects, EC awareness campaigns and training, and energy use monitoring. These functions accord closely with the requirement of the ECL (see Chapter 3).

Furthermore, the administrative structure for EC in public institutions extends to urban districts. For example, in 2013, in Huangdao Economic and Technological Development Zone, one of the three core urban districts in Qingdao, the Management Committee (the equivalent of an urban district authority) had a two-person Section of Energy Conservation Management, first established in 2011. It was responsible for EC work in 30-plus public institutions within the zone.[7] A similar structure at the urban district was found in Qingan District, Shanghai, in 2017. The district authority had a dozen sections under the district DRC, of which the Section of Environment and Resources was focused on *jieneng jianpai*. As of 2017, this section was led by a deputy director of the district DRC and had two full-time officers. However, only a third of urban district authorities in Shanghai reportedly had a dedicated section focused on environment and resources. Other authorities housed this function within sections focused on either industrial development or economic operation.[8]

Finally, the three cities also developed additional institutional capacities around MLCT outside their municipal DRC or GOA. For example, as of 2015, although only about ten people within the municipal DRC were working on *jieneng jianpai*-related issues in Shanghai, several organizations performed various auxiliary functions supporting their work. For example, the Low Carbon Economy and Climate Change Center (LCECCC) under Shanghai Information Center – a public institution affiliated to the municipal DRC – had a staff of 20-plus people as of March 2015. Similarly, the Shanghai Energy Conservation and Emissions Reduction Center (SECERC), established in June 2010 as a fully state-owned enterprise (SOE), had grown from a 3–4 person consultancy to a company with a staff of about 60 people by 2015.[9]

In addition, attention to coordination between government departments has increased, especially from the 13th FYP. Both Shanghai's Comprehensive Work Plan (CWP) for Emissions Reduction and Controlling GHG Emissions for the 13th FYP period (SMPG 2018) and Qingdao's Work Plan on

---

[7]   Interview 27 March 2013.
[8]   Interview 12 April 2017.
[9]   Interview 27 April 2015.

Controlling GHG Emissions (WPCGHGE) (QMPG 2018a) designate specific departments to take the lead for each of the specified tasks.

Compiling Local GHG Inventories and Monitoring Energy Use

The cities have started to develop knowledge about the sources of their emissions by compiling local GHG inventories, although this is not yet fully introduced nationwide. As one of the seven national pilots, Zhejiang is ahead in this area. The work is overseen by the Zhejiang Center for Climate Change and Low-Carbon Development Cooperation (ZCCCLCDC), established in June 2012 under the provincial DRC. By the end of 2015, Zhejiang had completed the inventories at three levels covering the province, 11 prefecture-level cities, and 90 county-level jurisdictions including urban districts. Moreover, there was only a one-year lag in the compilation, which means that inventories are updated annually. This work has become routinized since 2016 (ZCCCLCDC 2015). In addition, during 2014–15, the ZCCCLCDC developed and published 'Guidelines for the Preparation of Municipal and County-level Greenhouse Gas Inventories', the first of their kind in the country. It has also run a series of workshops and seminars to train specialists for this work (ZCCCLCDC 2015). As a result, Zhejiang has developed a comprehensive understanding of its annual emissions since 2005 at the provincial level and annual emissions by all the 11 municipalities after 2010. This covers the total amount of GHGs emissions,[10] composition, and distribution among key industries, enterprises and county-level jurisdictions. Key sources of emissions are continuously monitored (Gu 2016). In terms of application, the inventories for the 11 municipalities were used to aid the implementation of the GDP Carbon Intensity Reduction Target Responsibility System in 2014 for the first time.

Moreover, all three cities have conducted energy and carbon profiling for their MEUUs. During the 12th FYP period, Shanghai implemented energy audits for 646 MEUUs in industry and transport and 855 large-scale buildings (SMPG 2017a). Qingdao set up dedicated energy performance posts in 117 MEUUs, which are required to report energy use status regularly. Eighty-two MEUUs introduced an energy performance system (QMPG 2016b). By 2016, Hangzhou had completed carbon inventories for more than 130 key enterprises (Cheng and Guo 2016).

---

[10]   The inventory includes six types of GHG: CO_2, CH_4, N_2O, HFCs, PFCs, SF_6.

Developing Low-Carbon Strategies and Road Maps

Both Qingdao and Hangzhou developed LCT strategies within their juris-dictions before becoming LCCPs. Hangzhou was the first Chinese city to voluntarily commit itself to becoming a low-carbon city in 2008. It made this decision formal in 2009 and set out its actions in a 50-clause plan (HCCPMC and HMPG 2009). It also developed a low-carbon city development plan for the 12th FYP period (HMPG 2011) and a longer-term Climate Change Plan (2013–20) (HDRC 2014). Shanghai has formulated a '13th FYP for Energy Conservation and Climate Change' (SMPG 2017a) and a '13th FYP for Energy Development' as part of the FYP system (SMPG 2017b).

Qingdao faced especially grave challenges in making the LCT. This was partly because the municipal government introduced an economic develop-ment strategy of 'big industry, big development' in the 11th FYP (2006–10), just at a time when the central government started to emphasize *jieneng jianpai*. This strategy stimulated the growth of high-value-add, but energy- and emission-intensive, manufacturing activities. Thus, the secondary sector accounted for 57.7 per cent of Qingdao's total energy consumption and 55.7 per cent of its CO_2 emissions in 2008 (WRI 2011a); its share of energy con-sumption was still as high as 53.7 per cent, 52.9 per cent and 55.1 per cent in 2010, 2011 and 2012 respectively (Wang et al. 2016, p. 94). On the other hand, Qingdao has been pursuing aggressive territorial expansion since the early 1990s under the strategy of the 'Rim of Jiaozhou Bay Development'. This strategy is motivated by both the logic of land finance (see Xu 2019 for an introduction) and the objective of gaining international city status (Xue and Zhang 2014). In spatial terms, this strategy has transformed Qingdao from a relatively compact, monocentric city to a city with three sub-centres across Jiaozhou Bay. Consequently, the built-up areas expanded from 119.1 km^2 in 2000 to 291.5 km^2 in 2011, with significant adverse effects on for transport and related emissions (Xue and Zhang 2014). Building-related emissions have also grown rapidly, especially in the public sector. While CO_2 emissions of residen-tial buildings grew 5.2 per cent during 2005–9, those of public buildings grew by 104.3 per cent (WRI 2011b).

Nevertheless, drawing on the findings of the ADB-funded study (WRI 2012), Qingdao developed not only a detailed 'Qingdao City Low-Carbon Development Plan (2014–2020)' (hereafter 'Qingdao LCDP') (QDRC 2014), but also sectoral plans such as 'Special planning for clean energy heating in Qingdao (2014–2020)' (C40 Cities 2017). Qingdao's LCDP distinguishes the near-term focus on energy and economic restructuring from longer-term focus on transport and buildings. It identifies actions across five areas: industrial structure adjustment, energy conservation and efficiency improvement, devel-opment of non-fossil fuels, development of low-carbon transport, and increased

carbon sinks of forest. In addition, it sets out a number of performance indicators for 2020, as shown in Table 6.2. The latter shows that Qingdao aimed to
outperform the upper end of China's Cancun Pledge in terms of carbon intensity reductions, to increase the contribution of tertiary GDP from 52 per cent
to 58 per cent and to increase the forestry coverage ratio from 37 per cent to 45
per cent from 2010–20. However, its targets for energy intensity reduction and
non-fossil fuels development were below national targets.

Implementing the Energy Conservation and Emissions Reduction Target Responsibility Systems (ECERTRS)

The centrally administered ECERTRS is fully operative locally in the three
cities. As indicated in Chapter 4, this system comprises three interlocking
elements: setting of binding targets; monitoring, assessment and evaluation;
and sanctions. A comparison of Table 6.2 and Table 4.1 shows that Qingdao's
LCDP set more ambitious goals for itself than the national goals in some,
but not all areas. However, the city's mitigation commitments and ambition
expanded in subsequent years, to some extent driven by central pressure.
Qingao's WPCGHGE for the 13th FYP (QMPG 2018a), issued in March
2018, included a new indicator and set the proportion of green buildings as
a percentage of all newly-built at 50 per cent. Furthermore, the Qingdao Action
Programme for Energy Conservation and Green Development (2018–2020)
(QMPG 2018b), issued in August 2018, introduced yet more additional targets
including: (1) a total energy consumption cap of 33.8 million tce (2020); (2)
reduction of energy consumption per unit of area and per person (2015–20) for
public institutions by 10 per cent and 11 per cent respectively; (3) reduction of
individual consumption-related CO_2 emissions by 20 per cent from the level of
2015; and (4) an increase in the proportion of green buildings of all newly-built
from 50 per cent to 100 per cent. While the first reflects the 'double control'
mechanism discussed in Chapter 4, the third indicator marks the beginning
of an effort to reduce individual consumption-related emissions, unseen in
Shanghai and Hangzhou.

Nevertheless, setting and accepting planning targets appears to be a matter
of some negotiation and consultation between different levels of government.
In the case of Qingdao as an LCCP, it was agreed with the NDRC that the city
would be allowed to maintain a stable level of emissions at the peak levels for
10 years after peaking in 2020.[11] A similar situation appears to have ocurred
in Shanghai. As late as May 2015, when I was staying in the city, Shanghai
was still undecided whether it wanted to commit itself to peak its emissions

---

[11]   Interview 19 September 2018.

Table 6.2　Key low-carbon development indicators, Qingdao

Indicator		Unit	2010	2015	2020	Type*
1. General	GDP carbon intensity (excl. imported power)	t-CO$_2$/RMB10,000 (2010 price)	1.56	1.25	0.97	B
	GDP carbon intensity relative to 2010 level (excl. imported power)	%	–	20	38	B
	GDP carbon intensity relative to 2005 level (excl. imported power)	%	20	36	50	B
2. Industrial restructuring	Share of tertiary sector of GDP	%	46	52	58	A
	Share of strategic emerging industries of GDP	%	6	15	25	A
3. Energy conservation and efficiency improvement	GDP energy intensity relative to 2010 levels	%	–	19	32	B
	Added-value energy intensity of above-scale industrial units relative to 2010 levels	%	–	20	35	A
4. Non-fossil fuel development	Share of primary energy consumption	%	1	3	8	B
5. Low-carbon transportation	Share of public transport in total motorized travel	%	31	40	60	A
6. Increase in forest carbon sinks	Ratio of forest coverage	%	37	40	45	B
	Forest cubic volume	10,000 m^3	950	1082	1200	A

Note:　*A = anticipatory; B = binding. See Chapter 3 for an explanation of the difference between the two.
Source:　QDRC (2014, p. 19).

by 2025 or 2030 (i.e., the national target). Indeed, Shanghai's draft Master Plan (2016–40), issued in August 2016, still sets the CO_2 emissions peak year as 2025 (SMPG 2016b). The experiences of Qingdao and Shanghai appear to show that local authorities are nudged to adopt more ambitious ECER targets by their upper governments, specifically to set an earlier peaking year than the local authorities originally envisaged.

In order to implement their agreed planning targets, all three urban authorities undertook targets disaggregation among their urban districts. However, these cities adopted different approaches to this allocation. In Qingdao, this was done for districts/counties, sectors and MEUUs by the ECO under the DRC. All 10 urban districts were required to meet the same level of energy intensity reduction (16 per cent over the period of the 13th FYP (2015–20) and 2.5 per cent for 2018), identical to the planned target for the entire city (QMPG 2018a, Annex 1). In line with the 'double control' mechanism and echoing the inter-province system (see Table 4.2), a target was further set for each urban district in terms of the amount of energy consumption increase that can be permitted over the period, totalling 2.86 million tce. The allocation of this amount was, however, very uneven: Huangdao district received a quota nine times as much as that of the district with the smallest quota, likely reflective of the size of their respective energy consumption. Moreover, target responsibility contracts were signed between the heads of the urban districts and management teams of MEUUs on one side and the municipal DRC on the other side. This covered all enterprises consuming 3,000 tce or more per year (QDRC 2014, p. 31). Performance in meeting the contracted targets carries either incentives or penalties. Incentives can take different forms, ranging from financial rewards or accolades (for public sector users) to power cuts for enterprises.[12] Qingdao established a scheme in 2011, under which enterprises that perform well in energy conservation technological reform are rewarded with prize money and publicity. For example, in 2014, 29 projects belonging to 25 enterprises received prize money.

Shanghai also disaggregated its targets, by both sector and district/county during the 12th FYP. It set four main mitigation targets for the 13th FYP period: total energy consumption increase limited to 9.7 million tce; CO_2 emission for 2020 capped at 250 million tonnes; and energy intensity and CO_2 intensity of GDP to fall by 17 per cent and 20.5 per cent, respectively, in 2020 compared with the levels of 2015 (SMPG 2017a). The CO_2 cap is consistent with its pledge to peak its emissions by 2020. Its allocation of the targets is almost even across three classes of the urban districts despite significant differences in their existing energy and carbon intensities: the district target

---

[12]   Interview 28 March 2013.

for energy intensity reduction for the period ranges from 15 per cent to 17 per cent, whereas the district target for CO_2 intensity reduction for the same period ranges from 15 per cent to 18 per cent. Finally, all the urban districts were permitted no more than a 2 per cent increase in their energy consumption during the 13th FYP period. Shanghai's more egalitarian approach is not without its critics. Some district-level officials complained about the arbitrary nature of such allocations.[13] In contrast, in order to meet Hangzhou's overall mitigation target for the 13th FYP (2016–20) – a 25 per cent reduction of carbon intensity – the municipal government set very different mitigation targets for its various urban districts, ranging from 46 per cent for Gongshu district to 8 per cent for Chun-An county (HMPG 2017). This may reflect greater knowledge of emission sources by HMPG.

Monitoring has become routinized through energy use audits (*nengyuan shenji*), performed by specialized service providers. Meanwhile, the responsible city or district level administrative offices are tasked with scrutinizing the audit results and incorporate these into their annual reporting. Thus the Section of Environment and Resources in Qingan, Shanghai, followed a strict *jieneng jianpai* annual work cycle as follows: (1) beginning of the year: submission of annual plan on a template provided by the Municipal DRC; (2) quarterly: monitoring of the implementation of annual plan; (3) end of year: submission of annual report; (4) 1st quarter of the following year: auditing of the annual report by superior government department.[14] In Zhejiang, the ZCCCLCDC follows a similar annual cycle with key milestones in May, October, November and December for compiling GHG inventories (ZCCCLCDC 2015).

The accountability for the attainment of the targets is enforced on the basis of this monitoring. There is evidence that this accountability is binding. Qingdao DRC reported in 2016 that it successfully passed the province's ECERTRS evaluation for 2015 and the 12th FYP period. The report stated that during the 12th FYP period, a total of 382 inspection and monitoring activities (i.e., 1.5 activities per week) took place and that annual assessment of the districts' and sectors' target attainment was undertaken to ensure the fulfilment of the five-year targets (QMPG 2016b). It also indicated that province-controlled energy use statistics were utilized to verify the levels of fulfilment. Local officials in Shanghai reported that a failure to meet targets would disqualify the government section that they work for from winning any prizes, and that it would also deprive the officers involved of their annual bonus. Businesses would be denied power supply or closed down if failing to meet targets, as happened in the 11th FYP round in Qingdao. Like NDRC's annual publication

---

[13] Interview 12 April 2017.
[14] Interview 12 April 2017.

of provincial governments' performance in achieving their *jieneng jianpai* targets in the previous year, Shanghai publishes annual status reports on urban districts' performance in meeting their *jieneng jianpai* targets. A 2019 report shows that nine of Shanghai's 16 urban districts were rated as 'excellent', whereas the rest were rated as 'good'. Three indicators were used to rate the annual performance: decrease in energy intensity; growth of total energy consumption; and decrease in carbon intensity (SDRC 2019). Notably, these indicators are consistent with the 'double control' mechanism (cf. Table 4.2).

Industrial Policy

A common strategy by the cities is divestment from the so-called 'two highs and one surplus' industries (cf. Chapter 4). This is implemented through detailed industrial policy published by the city authorities by adapting national industrial policy. We will focus on Shanghai's experience here. Part of Shanghai's difficulty is that all the 10 top sources of emissions in Shanghai were centrally administered SOEs, over which Shanghai has little control. So it was a matter of negotiating with the central government about 'what Shanghai can stop producing'.[15] Nevertheless, sources close to the government suggested that productive capacities worth hundreds of million yuan belonging to the Jingshan Petrochemical Plant, a centrally controlled SOE, were written off on account of high emissions. For comparison, during the 12th FYP period, a total of 4,296 projects were closed down, reducing energy consumption by 4.39 million tce (SMPG 2017a). During the same period, Hangzhou phased out over 1,500 enterprises of this kind (Cheng and Guo 2016).

This process intensified in 2014 in Shanghai, involving coordinated actions by several municipal organs. First, in February, the Shanghai Development and Reform Commission (SDRC), Shanghai Economy and Information Technology Commission (SHETIC), and Shanghai Municipal Bureau of Finance (SMBF) issued a set of measures introducing differential power prices for the purpose of promoting the adjustment of industrial structure (SDRC et al. 2014). Then in March, the SHETIC (2014) published the Guiding Catalogue for Production Technologies, Equipment and Products of Restricted and Eliminated Industries in Shanghai (first batch). Finally in May, the SHETIC, Shanghai Municipal Bureau of Statistics (SMBS), Shanghai Municipal Energy Conservation Supervision Centre and Shanghai Municipal Energy Efficiency Center jointly issued a negative list for the adjustment of industrial structure and guidelines based on energy efficiency (SHETIC et al. 2014). Since then, these guiding catalogues and the negative list have been regularly updated by

---

[15] Interview 28 April 2015.

the SHEITC in cooperation with other municipal authorities. This process has been characterized by the expanding scope of coverage and rising standards. For example, while the numbers of items listed under the 'restricted' and 'eliminated' types were 196 and 59 items respectively in the first batch (SHETIC 2014), the Guiding Catalogue for the Adjustment of Industrial Structure (2020 edition) by SHEITC (2020) covers 435 and 346 items, respectively.

Experimental Piloting

There are widespread practices of piloting. Since all three cities are on the national LCCP programme (cf. Chapter 4), we will focus on exploring the cities' perspectives and actions here. The first thing to note is that pilots sometimes result from active lobbying from below. Hangzhou became one of the eight first-batch pilot jurisdictions in 2010, because it actively applied to the NDRC to conduct low-carbon experimentation work after its decision in 2009 to develop itself into a low-carbon city. It further showed its enthusiasm by offering rewards to low-carbon pilots within its jurisdiction: a prize of RMB 1 million (or RMB 500,000) was promised to any organization and its relevant local government that would lead a central-level low-carbon development pilot (or base) (HCCPMC and HMPG 2009).

There is evidence that piloting has cumulative effects as it brings three key benefits. These range from the financial and reputational, to learning and strengthened institutional capacity building. For example, after being designated as a first-batch LCCP in 2010, Hangzhou then became one of the comprehensive demonstration cities of national *jieneng jianpai* fiscal policy, launched by the MoF and NDRC in 2011. This would have enabled Hangzhou to receive an annual reward of RMB 400 million over three years (i.e., a total of RMB 1.2 billion over 2012–14) (Tanpaifang 2013). In May 2010, Qingdao was approved to be a RE demonstration city by the MoF and the Ministry of Housing and Urban and Rural Development (MHURD). By March 2011, there were 18 national-level demonstration projects within Qingdao, which together attracted RMB 155 million of state investment leveraging a total of RMB 5 billion for building-related RE projects (WRI 2011a, p. 45). On the other hand, various pilots are eligible for potential funding from the national Energy Conservation and Emissions Reduction Special Fund (ECERSF), or related sectoral funds.

Piloting also brings learning opportunities and reputational benefits. One of the learning opportunities that city officials genuinely value seems to involve interaction with superior authorities, peers, and international actors and organizations. For example, local officials and researchers in all three cities spoke of 'advanced concepts' (*xianjin linian* in Chinese) that international partners could bring. In another example, the Energy Foundation China provided advice to four out of the seven original carbon emissions pilots, including Shanghai (EFC 2019). China's success in promoting green finance is built upon its collaboration with international partners (Elliott and Zhang 2019). There are also reputational benefits: all the three case cities here are profiled in a book on the LCCP (GLCDTTP 2016). Qingdao and Shanghai also feature prominently in publications by the C40 (C40 and LBNL 2018) and the Asian Development Bank (ADB 2018).

Competitive piloting is also deployed at the sub-city level. Hangzhou's initial decision to turn itself into a low-carbon city called for the development of half a dozen pilots across different levels including community, enterprise, building, industrial park, new town and urban complex (HCCPMC and HMPG 2009). In June 2019, the QDRC issued a special 'Plan for Piloting Low-Carbon Communities, Towns and Industrial Zones'. It proposed that between 2019 and 2025, Qingdao would select eight to ten communities, five towns and four industrial zones to conduct low-carbon development pilots. By the end of 2025, 30 per cent of communities should become low-carbon, with five to ten communities to be recognized as national-level low-carbon demonstration communities (QDRC 2019).

However, becoming a pilot does not guarantee positive outcomes. For instance, Qingdao gained approval from the NDRC to pilot carbon emissions trading in November 2014 (Wang et al. 2016). However, this has not materialized due to a lack of response from other departments of the municipal authority.[16] This illustrates the importance of local institutional capacity and coordination in taking advantage of the opportunities that come with MLCT-related pilots.

Experimenting with Market-Based Mechanisms

New market-based mechanisms are being introduced for *jieneng jianpai*. Indeed, Jiang Zhao-Li, deputy director of the Department of Climate Change, NDRC, insists that exploring the pathways towards LCT through market-based mechanisms is a key objective of the LCCP programme (21st CBH 2016). Both Shanghai and Qingdao have made use of Energy Performance Contracting (EPC). EPC was initiated and trialled in Shanghai, before it gained the support of the NDRC and spread to the rest of the country. Qingdao issued implementing rules for rewarding EPC through prizes, rather than subsidies, from 2011. The standard of prize monies was RMB 300 per ton of coal saved. This sum includes RMB 240/tce from the central budget and RMB 60/tce from the municipal budget. During the 12th FYP period, in Qingdao, EPC helped reduce

---

[16] Interview 19 September 2018.

emissions by 102,000 tce, compared with a saving of 367,200 tce from energy technology upgrading (QMPG 2016b). Shanghai also provided subsidies for EPC, but in a different way. For example, depending on the extent of energy saving achieved and as of 2016, subsidies of between 20 and 30 per cent of capital investment costs were available in Jingan District (JDDRC 2016).

The other major experiment with market-based mechanism has been CET. This was an area where the centre provided scant guidance and simply encouraged the pilots to explore. Of the three cities, Shanghai is the only one that runs a CET pilot designated by the NDRC. After two years' preparation, the market started to trade on 26 November 2013. It included 191 MEUUs. The threshold was an annual emissions level of 20,000 t-CO_2 for industrial enterprises and 10,000 t-CO_2 for others (including aviation, airports, ports, supermarkets, hotels, commercial office buildings and railway stations) (Wu et al. 2016, p. 163). These enterprises accounted for about 60 per cent of the city's total emissions (ADB 2018). By the end of 2018, a cumulative volume of 1.06 gigatons of CO_2 emissions had been traded, involving a cumulative transaction value of RMB 1.108 billion (SEEE 2019, p. 7). Shanghai's achievements include a 100 per cent compliance ratio of the covered enterprises over five years in a row, being the first to initiate certified emissions reduction (CER) trading and having the largest number of institutional investors. However, in terms of the size of trading volumes, Shanghai's is not the biggest of the nine carbon trading pilots in China. By the end of March 2019, the total volume of transactions nationwide was 2.97 gigatons, involving RMB 6.487 billion, with highest trade volumes made by Guangdong and Hubei. Shenzhen, Shanghai and Beijing were considered as second-tier pilots (Tanpaifang 2019). Nevertheless, partially thanks to its status as China's premium financial centre, Shanghai has successfully positioned itself to play a central role in China's emerging national carbon emissions market. In May 2016, it won approval by the NDRC to set up the National Carbon Market Capacity Building (Shanghai) Center. Furthermore, in December 2017, the NDRC designated Shanghai to build the national carbon trading system (SEEE 2019). A representative of SEEE, whom I interviewed, indicated that a major benefit of managing the pilot was to raise awareness about carbon accounting and to help develop capacities for carbon management among the MEUUs.[17]

Population-Based Strategy and Land-Use Planning

There are contrasting strategies towards the management of population inflows in the context of LCT, either actively encouraging population growth or trying

---

[17]   Interview 21 April 2015.

to curb it. The three cities are in three different situations. With low population density and relatively slow population growth, population inflow, especially of degree holders, is actively encouraged by Qingdao and Hangzhou. Indeed, a local official suggested that recent reduction in emissions per capita in Qingdao was partly thanks to population growth.[18] In contrast, Shanghai municipal government has decided to cap its target population at the current level in its draft master plan (2016–40). Although this appears to be mainly motivated by the desire to improve the quality of life for its extant population, this strategy is nonetheless consistent with controlling its total emissions, thus helping it to meet its pledge to peak its emission by 2020.

There is also land-use-related adjustment. While Qingdao's 2006–20 Master Plan (QMPB 2009) reserved Hongdao in the north coast of Jiazhou Bay as the future administrative centre for the municipality, this idea had been quietly abandoned by the time of my visit in 2013.[19] On the other hand, Shanghai has started to think about densifying some areas. It also started to measure the performance of its various development zones in terms of output per unit of land area (SMLGCCECER 2019). Land-use planning has also changed: Shanghai has consolidated its manufacturing industries into some 200 sites and de-emphasized certain industries such as recycling in Linggan New Area well ahead of national policy change.[20]

International Collaboration

There is significant interest in and success of collaborating with international actors across the three cities. International collaboration has taken a variety of forms. For example, the ADB provided technical assistance for Shanghai's CET pilot, a grant for the development of the low-carbon roadmap for Qingdao (WRI 2012) and a USD 130 million loan to Qingdao for a district heating and cooling system powered by air-, ground-, and waste-source heat pumps (ADB 2018). The WRI is active in these cities too. It coordinated the development of Qingdao's LCDP and helped Zhejiang province to develop its three-level system of GHG inventory compilation during 2014–15, with funding from the Ministry of Foreign Affairs of the Netherlands (ZCCCLCDC 2015). After joining the C40 in 2017, Qingdao's clean energy building programme was one of the four finalists in a competition called 2017 Award Cities4Energy (C40 Cities 2017). Qingdao is also one of the four participating cities in the C40's China Buildings Programme (2017–20) (C40 and LBNL 2018).

---

[18] Interview 19 September 2018.
[19] Interview 24 March 2013.
[20] Interviews 12 May 2015 and 8 May 2015.

Interviews in Qingdao and media reports reveal some shortfall in fulfilling local officials' ambition to pioneer a new LCT model by working with transnational corporations (TNCs). At one point, the city's DRC hoped to establish three separate demonstration zones respectively with American, European and Japanese commercial partners. Indeed, in 2010 and 2011, it signed memorandums of understanding (MoUs) with GE (America), Suez Environment (French), Mizuho (Japan), and Siemens (Germany) for this purpose amidst significant publicity (Zhou 2010; WRI 2011c). However, only the cooperation with GDF Suez has borne fruit. In October 2012, an agreement was signed with the French-owned company regarding port development, wastewater processing and comprehensive use in order to promote a low-carbon, circular economy in Qingdao West Coast Development Zone (Water 8848 2014). The project with Suez Environment was a 50:50 joint venture, with a total investment of RMB 500 million for the first phase. The other projects failed to proceed for a mix of reasons including stringent requirement for infrastructural and institutional support, and a lack of assured returns.[21]

Awareness Raising Campaigns

Awareness for *jieneng jianpai* work is being raised across society. Local officials that I met spoke knowledgeably about *jieneng jianpai* activities. For example, an official in Qingdao's GOA highlighted during an interview[22] that relevant national regulation stipulates that in offices of public institutions using air conditioners, room temperature must not be lower than 26°C in summer or higher than 20°C in winter; they also had rules regarding shutting down computers at the end of the working day. A variety of means is deployed by the GOA in Qingdao for raising awareness, including essay competitions, poster competitions, school curricular developments and the Energy Conservation Awareness Week (11–17 June).[23] The Energy and Environment Section of Qingan district authority, Shanghai, had an 'energy conservation gold idea' programme operating through the union.[24]

However, arguably the most successful awareness campaign is Ant Forest, a digital app. Launched in August 2016, it was developed by Ant Financial, a fin-tech company based in Hangzhou and a subsidiary of Alibaba. Essentially, the app tracks multiple aspects of users' daily lives through integration with Alipay and Alibaba. It then rewards users for their sustainable behaviours with

---

21   Interview 24 March 2013.
22   Interview 28 March 2013.
23   Interview 28 March 2013.
24   Interview 12 April 2017.

green energy value (GEV), which can be used to grow initially 'virtual trees' and eventually real trees in Inner Mongolia to fight desertification there (ADB 2018). A study by the Policy Research Center on Environment and Economy (PRCEE 2019) under the Ministry of Ecology and Environment shows that the app also allows the users to adopt an area for protection in the Deqin Public Welfare Reserve, Yunnan province, in southwest China.

There are three key links in the operation of this app:

- First, Alibaba and China Beijing Environmental Exchange (CBEEX) worked together to produce green and low-carbon scenarios and endow each behaviour with recordable and quantifiable GEV (1 g of GEV is equivalent to 1 g of carbon emission saved). The targeted activities cover five broad categories: green-mode travel; reduced travel; reduced use of paper and plastics; improved energy efficiency; and recycling. The categories in turn encompass 17 concrete behaviours such as walking, substitution of plastic bags and cookery, online ticketing, electronic sub-mission, and the like. A defined amount of GEV is required for a particular outcome. For example, 2,700 g of GEV earned can be exchanged for the protection of one square metre of the Reserve in Yunnan; alternatively, having a real red willow or Populus euphratica tree planted requires 22,400 g or 215,680 g of GEV earned.
- Second, individual users are encouraged to adopt these sustainable behav-iours through gamification. For example, friends, family members, fan club members, or university alumni can combine their GEVs for growing trees or protecting a part of the Reserve.
- Third, in collaboration with charitable organizations including the Alibaba Foundation, Ant Finance donates monies for real, identifiable trees to be planted in designated areas by NGOs. In addition, as partner organizations, companies such as the Gago Group provide technological solutions to enable users to 'see' their trees and to appreciate their trees' environmental impact via satellite images.

According to the report, by the end of August 2019 and at its third anniversary, the Ant Forest app had attracted more than 500 million users. It had led to the planting and upkeep of 122 million trees covering an area as big as Hong Kong and secured protection for 15,000 hectares of the reserve land in Yunnan. Moreover, this app has resulted in the reduction of 7.92 million tonnes of CO_2 and provided employment opportunities for 400,000 people. Interestingly, Hangzhou and Shanghai took the first and second spots in a league of cities according to the amount of emissions reduced through walking between August 2018 and August 2019, whereas the fastest growth was registered in western inland regions (PRCEE 2019).

It is possible to find Ant Forest's roots in Hangzhou's pioneering stance on and practice of low-carbon development. The city's 'six-in-one' low-carbon development goal includes lifestyle and society as two of the six foci. Since 2008, Hangzhou has built the world's most comprehensive shared bicycle system. As early as 2011, it developed low-carbon community standards and evaluative benchmarks. Residents of some communities were rewarded with cash points or household goods for waste recycling. In 2015, a resident was able to win a five-year use-right of a BYD hybrid electric vehicle. To stimulate public participation, the city carried out family low-carbon actions among 100,000 households (Liu et al. 2015).

FINANCING AT CITY LEVEL

Despite the diverse rationalities and GTTs, the cities have all tried to attract as much capital as possible for MLCT activities. They have developed ambitious investment plans. Hangzhou's Climate Change Plan (2013–20) states that during the eight-year period, for the purpose of addressing climate change, 235 projects will be completed, involving a total investment of RMB 442.234 billion across seven areas (HDRC 2014). Averaging this sum over eight years suggests RMB 55.279 billion per annum, the equivalent of just over 10 per cent of Hangzhou's annual FAI, or 5.1 per cent of GDP during 2013–18.[25] Interestingly, this investment package represents a eight-fold increase of the investment programme proposed only three years ago in 2011, when Hangzhou's 12th FYP for Low-Carbon Development (HMPG 2011) proposed to develop 51 low-carbon demonstration projects with a total investment of RMB 55.342 billion, of which investments worth RMB 37.196 billion would be completed during the 12th FYP. In the case of Qingdao, the WRI-led study estimates that, for achieving the 45 per cent carbon intensity reduction scenario and under various assumptions, additional investment needs from 2010–20 would be RMB 208 billion, including RMB 77 billion alone in the power sector. This suggests an annual investment need of RMB 20.8 billion (approximately 3.1 per cent of the FAI or 2.2 per cent of GDP in 2015). The study finds Qingdao's average mitigation cost will be RMB 240/t-CO_2, while its total mitigation costs amount to RMB 10.5 billion during the 10 years from 2010–20 (WRI 2012). Different sources of finance have been exploited.

---

[25]   The average FAI was RMB 549.254 bn over the period according to official statistical bulletins.

Traditional Sources

GPB allocation under the heading of *jieneng jianpai* and state-sanctioned bank loans are key sources of financing for all three cities. However, Hangzhou and Qingdao have not published detailed public spending figures at all, whereas Shanghai has only started publishing this figure from 2015 through its annual socio-economic development statistical bulletins. The latter show that Shanghai's energy conservation and environmental protection (ECEP) spending has increased from RMB 10.435 billion in 2015, to RMB 13.441 billion in 2016, RMB 22.460 billion in 2017, and RMB 23.339 billion in 2018, at 30.8 per cent CAGR over the period. Meanwhile, Shanghai's investment on environmental protection as a percentage of GDP has risen from 2.82 per cent in 2015 to 3.1 per cent in 2017 (SMBS 2018).

In addition, Shanghai has set up its own ECERSF. The scheme was first announced jointly by the municipal DRC and the Bureau of Finance in June 2008 and was updated in February 2017. Both editions indicate three main sources of funding: (1) allocation from general municipal-level budget; (2) income from the implementation of differential electricity tariffs; and (3) other sources determined by the municipal government. They identify a dozen categories of eligible project, but in slightly different orders of importance. The categories of the 2017 version are: (1) elimination of outdated productive capacities; (2) industrial *jieneng jianpai*; (3) EPC; (4) buildings *jieneng jianpai*; (5), transport *jieneng jianpai*; (6) RE and clean energy; (7) air pollution mitigation; (8) water pollution mitigation; (9) circular economy; (10) low-carbon development and addressing climate change; (11) energy-saving products promotion and management capacity building; and (12) co-financing for centrally designated items and others. For each category or sub-category, a relevant government department is designated as responsible for formulating the implementation rules and actual operationalization. A comparison of the two editions shows that low-carbon development and addressing climate change were newly added to the 2017 version.

The monies from Shanghai's ECERSF are dispensed in two principal forms: subsidies and prizes. Statistics from Shanghai's DRC website show that the amount dispensed from this fund was RMB 1.544 billion, RMB 1.387 billion, RMB 1.576 billion, RMB 2.243 billion, RMB 2.665 billion, RMB 3.509 billion, RMB 3.238 billion, respectively in 2010, 2011, 2012, 2013, 2014, 2015 and 2017.[26] This is equivalent to 13.1 per cent CAGR in nominal terms. Since a similar amount (in a 1:1 ratio) is believed to be spent by the district

---

[26] An overall figure for 2016 has not been found. Spendings on three sub-funds (circular economy and comprehensive resources usage; distributed energy supply and

authorities through co-funding, the total annual public spending is about double that amount (i.e., about RMB 6–7 billion). Moreover, the leverage ratio for private capital is estimated to be eight to ten times.[27] A Shanghai DRC official, whom I interviewed, also indicated that in addition to the spending through the ECERSF, as of 2015, annual infrastructure investment related to the LCT amounted to RMB 60–70 billion (approximately 10 per cent of the FAI in the city).[28] These figures suggest an annual spending in excess of RMB 120 billion (including private investment) (approximately 4.8 per cent of GDP).

Some resourceful district authorities have set up their own special funds for *jieneng jianpai*. For example, Qingan district, Shanghai, has operated such a fund since, reportedly, 2004.[29] It had an initial capital of RMB 200 million to help the district to meet its energy conservation targets by incentivizing the relocation of large emitters such as shipping companies. By 2017, the monies available from this fund amounted to about RMB 20–30 million per annum (about 0.2 per cent of the district authority's annual budget). Spending, either in subsidies or prize monies, was concentrated in two main areas: (1) retrofitting public sector buildings; (2) subsidizing enterprises' energy conservation endeavours especially if such actions are mandated by the government. The scheme valid from October 2016 identifies nine categories, in significant consistency with those of the municipal ECERSF. For example, the first category covers independent energy conservation and reform initiatives, where 20 per cent of total capital costs can be recovered from subsidies. Category 8 appears to cover large energy conservation efforts involving energy savings of at least 50 tce; a prize of RMB 600 is awarded for every tce saved. Finally, one-off prizes are provided for a range of good performance. For example, winning a municipal-level 'energy conservation frontrunner' award would attract a cash prize of RMB 60,000 (for gold) or RMB 20,000 (for silver) (JDDRC 2016). Matching funding was provided for PV projects. Here the subsidies were from three levels of government: central government, municipal (70 per cent of central subsidy), and district (35 per cent of central subsidy).[30]

Hangzhou has reportedly set up a low-carbon fund of RMB 5 billion (Wang et al. 2015). According to HMPGO (2013), this fund can be spent on prizes in lieu of subsidies, subsidies for loan interest payments and fiscal subsidies. The ratio of fiscal subsidies for corporate investment is up to 20 per cent, capped

---

gas-fired air conditioning; renewables and new energy development) add up to RMB 241.129 million.

[27]   Interview 28 April 2015.
[28]   Interview 28 April 2015.
[29]   Interview 12 April 2017.
[30]   Interview 12 April 2017.

at RMB 10 million per project. This ratio can be as high as 25 per cent for projects whose investment is led mainly by the government. Information on LCT funding in Qingdao is sparse. The city has limited financial resources in comparison with other major Chinese cities (WRI 2012). Bank loans, public subsidies, tax rebates and foreign loans are its principal sources of finance. For example, Qingdao DRC published detailed funding sources for three key projects on resource conservation and circular use in 2017 that were labelled as 'within the central budget'. The total investment involved was RMB 5.4509 billion. The contributions from the public budget, bank loans and self-raised capital were respectively 2.2 per cent, 66.7 per cent and 31.1 per cent (QDRC 2017). In another example, in July 2012, as one of the second-batch pilot cities for low-carbon transport networks by the Ministry of Transport, Qingdao was allowed to undertake a package of 33 projects with a total investment of RMB 18.67 billion during 2012–14. The investment drew from three sources: central government subsidies, the municipal government (about 40 per cent), and enterprises themselves. At a cost of RMB 13 billion, the single largest investment was the construction of a subway for the city.[31] According to the MoT (2012), such pilot projects can enjoy financial support from the national ECERSF for Transport (MoF and MoT 2011). The wastewater processing plant co-owned by Qingdao West Coast Development Corporation and Suez Environment received a subsidy of RMB 17.8 million from the central budget in 2015 (QWCDG 2017).

According to the ADB (2018), Qingdao has invested RMB 3.65 billion in RE systems and buildings retrofit since 2012, more than half of which came from public funds. The city is investing RMB 23.2 billion in a clean district heating network covering 180 km^2 that will make use of air-, ground-, and waste-source heat pumps. The latter benefited from technical assistance and a USD 130 million loan for the project from the ADB (ADB 2018, p. 7). According to local media, the ADB-supported project involved a total investment of RMB 1.6 billion (approximately USD 260 million); upon its completion in 2018, the project was able to provide an additional clean energy and RE capacity of 863 MW, about one-half of the planned clean heating supply capacities for 2020 (Wang 2014).

Like Shanghai, Qingdao has also offered multi-level subsidies. For example, the Qingdao government has invested quite heavily in retrofitting its older residential buildings stock. According to C40 and LBNL (2018), the city retrofitted 7.85 million m^2 during the 12th FYP and aimed to retrofit at least 10 million m^2 during the 13th FYP. In fact, between 2016 and mid-2018, the city retrofitted 11.3 million m^2 and was set to retrofit an additional 15 million

---

[31]   Interview in Qingdao, 26 March 2013.

m² of residential buildings during 2018–20. During the 12th FYP, households doing comprehensive retrofitting, which includes both building envelope and heat metering, received subsidies up to RMB 300/m², with the central government providing RMB 45/m² and the city and district authorities RMB 255/ m². Due to rising demand, however, this model was considered unsustainable. So the municipal government planned to shift from a subsidy-oriented model towards a mixed subsidy model with wider private participation during the 13th FYP. The level of subsidies for comprehensive retrofitting would be capped at a maximum of RMB 135/m² during the 13th FYP. It also intended to switch from up-front payment to post-completion result-based payment (C40 and LBNL 2018). One of Qingdao's financing innovations is that residents in the Jimo District were offered free power for 25 years in exchange for use of their roof space. Specifically, a company provided RMB 1 million to install PV panels on 72 residential roofs; each of the systems is capable of producing 4 kW a day, twice the average power consumption of each property. The company receives income from the surplus electricity sold back to the power grid (ADB 2018, p. 8).

Green Finance

The three cities have made active use of green finance. The Bank of Qingdao and the Shanghai Pudong Development Bank (SPDB) were among the first four Chinese banks to issue green bonds in 2016. In 2019, Shandong, Zhejiang and Shanghai raised respectively RMB 24.05 billion, RMB 23.31 billion, and RMB 4.39 billion from green bonds, claiming 18.3 per cent of the total amount of capital raised in the country through this means. The SPDB's green credits portfolio increased from RMB 25.66 billion in 2012 (Yu et al. 2013) to RMB 217.52 billion at the end of 2018, at 42.8 per cent CAGR (DRCNET 2020).

Shandong province is the first Chinese province that has set up a green fund with loans from international financial institutions. Established in June 2018, the Shandong Green Development Fund has capital of RMB 10 billion. It is financed by an EUR 88.73 million (USD 100 million) loan from the Green Climate Fund and is administered by the ADB. The loan has a maturity period of 20 years (with 15 years' grace period) with interest at the 6-month European Interbank Rate plus a floating spread (0.3 per cent ca). Seventy-five per cent of the fund will be used on mitigation, whereas the remaining 25 per cent will be for adaptation. It is estimated that, by 2027, this fund will help to reduce emissions by 2.5 million tonnes of CO_2 per annum and improve the adaptability of 2 million people (CFEN 2019).

Qingdao has done well in utilizing green bonds. In 2016, the Commercial Bank of Qingdao was allowed by the PBC to issue two batches of green bonds, raising a total of RMB 8 billion (Xinhua Finance 2016). By the end of the third quarter of 2019, apart from RMB 3.5 billion that had come to maturity, a total of RMB 4.5 billion was used for green loans supporting 103 projects for with industrial energy conservation (24 per cent) and water conservation and non-conventional water use (28 per cent) claiming a lion's share (Bank of Qingdao 2019). Moreover, the Rural Bank of Qingdao raised RMB 3 billion through green bonds issued between July 2017 and July 2018 (QRCB 2018).

OUTCOMES

In terms of mitigation performance, the three cities exhibit big differences, although their exact extents are difficult to establish due to the lack of strictly comparable data. Table 6.3 pulls together statistics from different sources and should be treated as indicative. The data in columns 2–4 are from a single source. These show that all three cities had rising levels of CO_2 emissions per capita during the 2005–11 period. Of the three cities, Shanghai appeared to have the best performance during this period, as both its emissions per capita in 2011 and their growth rate during 2005–11 were the lowest among the three. The data for the 12th FYP period (2011–15) are from official city sources. They show that Shanghai again managed to reduce its energy and carbon intensities the most, by 24.45 per cent and 28.58 per cent, respectively. Hangzhou's performance is only marginally behind Shanghai in these respects. Moreover, as of 2015, its energy intensity was lower than Shanghai's, while its share of tertiary sector GDP was 6 per cent lower than Shanghai's (Table 6.1). In fact, as home to the Internet giant Alibaba and a host of ICT-based businesses, Hangzhou has an economy that is least dependent on energy-intensive industries among the three.

CO_2 emissions data for the 13th FYP period (2016–20) are not yet available. Yet initial signs are that energy intensity continued to fall in all three cities during this period. In Shanghai, the energy intensity of GDP decreased from 0.44 tce/RMB 10,000 in 2016 to 0.35 tce/RMB 10,000 in 2018, a reduction of 20.5 per cent. In Hangzhou, the same indicator fell by 6.9 per cent, 3.4 per cent, 4.3 per cent and 5.0 per cent respectively in 2016, 2017, 2018 and 2019, totalling 19.6 per cent over the four-year period.

Qingdao has not published detailed data on energy use and its efficiency over the years. Qingdao's CO_2 per capita rose sharply during 2005–11, largely due to its 'big industry, big development' strategy. Indeed, Qingdao's emissions per capita grew nearly 12 times faster in comparison with Shanghai during the same period. Of the three cities, Qingdao has achieved the lowest reduction in energy intensity (Table 6.3). Qingdao DRC reports that the city's total estimated CO_2 emissions (excluding electricity imports) in 2017 amounted to 72.72 million tonnes, whereas its CO_2 emissions per unit of GDP totalled 0.67 tCO$_2$ per RMB 10,000 (QDRC 2018). In comparison, an earlier

Table 6.3   *Emissions and mitigation indicators, Shanghai, Qingdao, and Hangzhou*

City	CO$_2$ per capita (tonne/ per person) (2005)	CO$_2$ per capita (tonne/ per person) (2011)	Growth in CO$_2$ per capita (2005–11) (%)	Energy consumption per unit of GDP (tce/ RMB 10,000) (2015)	Reduction in energy consumption per unit of GDP (2011–15) (%)	Reduction in CO$_2$ emissions per unit of GDP (2011–15) (%)
Shanghai	12.4	13.1	5.9	0.463	25.45	28.58
Qingdao	8.9	15.1	70.2	0.68 (2012)	17.7 (2011–14)	18
Hangzhou	10.5	15.9	51.8	0.43	22.5 (est.)	21.9 (2011–14)

Source: Wang et al. (2015) for columns 2–4; SMPG (2017a); QDRC (2014), QMPG (2016a); HMPG (2016).

report (QDRC 2014) indicates that the city's energy-related CO_2 emissions were 57.46 million tonnes and 93.82 million tonnes respectively in 2005 and 2011, whereas CO_2 intensity of GDP and CO_2 per capita were 1.48 tCO_2/RMB 10,000 and 10.67 tonnes in 2011, respectively. Under assumption of comparability, these figures imply that Qingdao's CO_2 emissions per capita in 2017 amounted to 7.83 tonnes,[32] 26 per cent lower than its 2011 levels; moreover, its CO_2 intensity has fallen from 1.48 tCO_2/RMB 10,000 in 2011 to 0.67 tCO_2/RMB 10,000 in 2017, a fall of 54.7 per cent.[33] This again suggests accelerated mitigation during the 13th FYP. Thus, despite the obvious differences between the three cities and their heavy reliance on manufacturing, they have all made significant progress in mitigation since the early 2010s.

The experience of Qingdao illustrates most strongly both the severity of the LCT challenge for a city that relies heavily on its energy-intensive manufacturing activities and the progress that has been made. Earlier scenario analysis suggested that even if the target of 45 per cent CO_2 intensity reduction is met, Qingdao's total CO_2 emissions (excluding imported electricity) in 2020 would still reach 179.3 million tonnes, while its CO_2 emissions per capita will be 13.4 tonnes in 2015 and 17.4 tonnes in 2020 (WRI 2012). How then do we explain the fact that both the total emissions and CO_2 emissions per capita are so much lower in 2017 than anticipated by the study? It would seem that this resulted from a significant slow-down of its economic growth coupled with population growth. While the WRI-led study assumes that its GDP would grow annually by 12 per cent during the 12th FYP (2011–15) and 11 per cent during the 13th FYP (2016–20), the actual growth rate was only 9.7 per cent for 2011–15 and 7.6 per cent per annum over 2016–18 (see Table 6.1). In comparison, the official 12th FYP anticipated a GDP growth rate of 11 per cent per annum during 2011–15 (QMPG 2011).

There is no doubt that the industrial restructuring stimulated by MLCT has produced a marked dampening effect on economic growth in the cities. This is because the divestment and structural adjustment deprived the cities of some of their hitherto most powerful engines of growth, that is, energy-intensive manufacturing. In Shanghai, gross industrial output (GIO) grew by only 2.2 per cent per annum during 2011–19, compared with 13.4 per cent per annum over 2006–10. The GIO of the seven industrial zones within the Pudong New Area, supposedly the most dynamic part of the city economy, grew by an

---

[32]   This derives from dividing total emissions by the number of permanent residents living in Qingdao in 2017 (9.29 million). The emissions total excludes emissions associated with imported electricity, therefore underestimates the levels of emissions.

[33]   The two sets of data are from different sources and therefore not strictly comparable.

average rate of 2.62 per cent per year over the period 2012–17.[34] The performance of the Lingan Area is particularly disappointing, but also illustrative. Located 75 km away from the centre of Shanghai and with a planned area of 315 km², Lingan Area was designated in 2002 as an industrial functional area. It is strategically located 37 km from Pudong Airport and 32 km from the Yangshan Deep Water Port in Hangzhou Bay via the East Sea Bridge. With designated industrial zones of 241 km², the Lingan Area aimed to rejuvenate Shanghai's manufacturing and respond to the centre's call for accelerating the development of strategically important industries. Planners had envisaged developing the Lingan Industrial Zone into a 'smart manufacturing' district with high-end manufacturing activities. However, it has struggled to meet its key manufacturing ambitions. For instance, its GIO grew by 2.49 per cent, minus 2.0 per cent and minus 2.7 per cent respectively in 2014, 2016 and 2017, although it achieved a robust growth of 24.8 per cent in 2018 (SPNAPG 2021). The experience of Shanghai demonstrates again the challenge of maintaining high growth during the LCT, although its success in attracting Tesla, the EV maker, to Lingan in 2018 also indicates potential opportunities associated with the MLCT.

CONCLUSION

This chapter explored the MLCT experiences in Shanghai, Qingdao and Hangzhou and, to some extent, the two provinces to which Qingdao and Hangzhou belong. It reveals a tremendous level of local effort to do two things: to meet planned mitigation targets within the ECERTRS and to lay a foundation for a lower-carbon and more sustainable economy. This exploration suggests that these cities all face enormous mitigation pressures and have often been encouraged by their superior governments to do more from one plan to the next. Nevertheless, they have also demonstrated strong commitments to meeting the targets and have so far managed to out-perform their targets.

The cities are affected by a significant tension between heavy industry being a key driver of local economic growth and the most important source of emissions simultaneously. The response so far appears to have been at the expense of growth, driven mainly by administrative pressure, rather than market signals. This is because mitigation costs are a lot higher than the prevailing carbon prices. On the other hand, there is a growing gap between the current focus on controlling industrial emissions and the rapid growth of transport and building-related emissions, locked-in by a carbon-intensive infrastructure that

---

[34]   According to the Pudong Yearbook. Data for 2016 is unavailable, so this growth rate covers only five years.

has been developed to support a somewhat out-of-date economic development strategy revolving around export-oriented industrialization and over-sized city development. Cities are now trying to readjust their strategies. Future actions are likely to be focused on transport and buildings. On the positive side, there is incipient effort to regulate consumption-related emissions in both Hangzhou and Qingdao.

In terms of GTTs, the three cities have followed religiously the disaggregation method and the ECERTRS to ensure the fulfilment of their targets. In addition, they have adopted many other GTTs, some of which are traditional, including institutional capacity building, planning, industrial policy, piloting, and international collaboration. Meanwhile, new kinds of GTT have been developed, including regular compilation of detailed GHG inventories, energy use monitoring of MEUUs, economic instruments including differential power prices, EPC, carbon trading, green finance, and ICT-enabled awareness campaigns. These two types are often blended. For instance, industrial policy is coupled with differential power prices in Shanghai and this is the case also in Hangzhou (Cheng and Guo 2016). At the same time, a substantial sum, ranging from 2–5 per cent of GDP or 3–10 per cent of IFA, is being spent on incentivizing LCT activities and developing climate-compliant infrastructure. Multi-billion yuan investment programmes have been undertaken for MLCT.

7. Conclusion: the rise of a carbon governmental state in China

If the 120-strong governments who have proclaimed a goal of carbon neutrality by 2050s live up to their promises, we are on the cusp of a great wave of mitigation and low-carbon transitions (MLCT) across the world in the next three decades. Of course, only history can tell whether this is to be. As the Cancun Pledge expires and the implementation of the Paris Agreement commences, this is an opportune time for us to reflect on the lessons that we could learn from China's MLCT experience in the past 10 years, when it managed to achieve one of the fastest decarbonizations in the world. This is indeed one of the book's chief objectives. Over the previous six chapters and with the aid of an experimental governmentality lens, we have addressed two of the three key research questions raised in Chapter 1, relating to the evidence of China's progress, especially during the 2010s, and the roles of political rationalities and governing techniques and technologies (GTTs) in supporting this performance. It is now time for us to draw together our findings from individual chapters and to address the third question, on the lessons that other nation states could learn from China's experience. We will also reflect on China's own lessons. This chapter is in three parts. The first section summarizes the findings related to the first two questions. The second section evaluates the analytical purchases of the carbon governmentality model that I have applied. The final section reflects upon China's lessons.

A SUMMARY OF SUBSTANTIVE RESEARCH FINDINGS

Chapter 1 shows that, despite its status as the world's largest CO_2 emitter, China has made significant progress in MLCT since the early 2010s. This performance is evident not only in the decoupling of its CO_2 emissions and CO_2 emissions per capita increases from income rises, but also in China's tremendous investment in and deployment of low-carbon technologies and the fostering of its large renewable energy (RE) equipment manufacturing capacities. In addition, it is manifested in China's effectiveness in fulfilling its Cancun Pledge and its continuous efforts to raise its climate ambition.

Chapter 2 explored existing literature on Chinese MLCT in a search for an explanation of China's good performance. Three models – the policy, politics and governance models – and their applications were critically reviewed. We showed that none of these models has provided or could provide satisfactory explanation for China's recent progress, albeit for different reasons. To summarize, the 'policy model' is unsatisfactory because it places a strong emphasis primarily on carbon pricing (Stern 2007), whereas China is evidently far from having established a meaningful carbon market despite its various marketization experiments. Neither does the 'politics model' work because of its analytical focus on the negative power of the incumbent actors and their resistance to change. Primarily through the lens of 'fragmented authoritarianism' or 'environmental authoritarianism', existing studies have focused on the problems of command and control (CAC) mechanisms, rather than aspects and factors that have driven MLCT forward in China. Finally, preoccupied with the role of non-state actors, rather than state actors, the 'governance model' (in the narrow sense, as explained in the chapter) is ineffectual for the case of China, where the state actors obviously have played a central role despite an incipient process of political pluralization. Overall, this review shows that extant studies have not paid sufficient attention to state actors and their actions, especially with regard to what makes them effective.

To make better sense of the Chinese experience, we then developed an alternative model – the model of carbon governmentality – in later parts of Chapter 2. This model conceptualizes MLCT as the outcome of an art of carbon government in the Foucauldian sense, treating government as the conduct of the conduct of the governed. Through the 'analytics of government' framework that Dean (2010) and others have developed, which we further adapted, Chapter 2 identified two key research areas, the relevant political rationalities and the GTTs by which authority is constituted and rule accomplished (see Table 2.1). Following this framework, we then explored the key elements of China's carbon governmentalities in Chapters 3–6.

We first examined policy makers' political rationalities for pursuing MLCT in Chapter 3. This covers who and what to be governed, the problems to be solved, the objectives to be achieved, forms of knowledge utilized and produced, the form of identities presupposed and the transformation sought. This investigation shows that, from the mid-2000s, the political elites have come to identify China's long-term development goals of modernization, national rejuvenation, and the ultimate triumph of Chinese socialism with the pursuit of green development including MLCT. This means that MLCT has been able to draw strong support from both pre-existing discourses about modernization and industrialization (pre-reform), the criterion of the truth (1978), the Four Cardinal Principles (1979), and the Socialist Market Economy (SME) (1992), and new discourses including the Scientific Outlook on Development (SOD)

(2007), Ecological Civilization (2012), and the Community of Common Destiny (CCD) (2012) on the international front. Moreover, MLCT is placed at the forefront in the search for a new economic development model and in constructing the Ecological Civilization, defined as a resource-efficient and environment-friendly society. It also led to a renewed appreciation of the intrinsic value of nature as assets and resources, a rebalancing of the human–nature relationship, and the reprioritization of performance criteria in the pre-existing cadre evaluation system (CES). This means that low-carbon rationality has gained solid support from a set of discourses that increasingly regard sustainable, green and low-carbon development as the only meaningful choice for the country and the socialist system. Part of the success of the collective discursive efforts also lies in the system's ability to permit intra-elite debates to reconnect with traditional cultural values about the harmony of humans and nature, and to draw on Chairman Xi's personal authority and his repeated message that 'green mountains and clear waters are invaluable assets'.

Chapter 3 also demonstrated the critical importance of law and regulations, that is, juridical power, in defining the key objects and subjects. There are eight key subject groups in China's carbon government: sub-national governments (SNGs) (including state-owned enterprises (SOEs)) and officials; energy-using units (EUUs) (especially MEUUs); manufacturers and suppliers of energy-consuming appliances and equipment (ECAE); RE investors; power suppliers and power grid companies; individuals as consumers and citizens; households; and finally, financial institutions (FIs). We found that, because of a prior 'management by region' governmentality for energy conservation (EC), territorial knowledge about sources of emissions, and existing regional institutional capacities, China's carbon governmentality has singled out SNGs, government officials and MEUUs managers as the most important subjects.

We then analysed the carbon GTTs developed and deployed to achieve MLCT objectives in Chapter 4. It was found that such GTTs are aimed at two principal sets of governing relationship: (1) centre vis-à-vis localities/ SOEs; (2) state vis-à-vis non-state firms and households/individuals. At the centre–locality interaction, the current system utilizes primarily juridical and disciplinary powers, supplemented by governmental power drawing on discourse and knowledge. It includes planning of various kinds and different lengths, regional/sectoral disaggregation, target responsibility systems (TRS), bureaucratic restructuring, re-prioritization of performance criteria for cadres and SOE managers, encouragement of experimentation, and institutional capacity building. However, in terms of ensuring the fulfilment of planning objectives, by far the most important mechanism is the Energy Conservation and Emissions Reduction Target Responsibility System (ECERTRS) comprising target setting via central planning and disaggregation, regular monitoring, assessment and evaluation (MA&E), and strong sanctions for irresponsible

SNGs, officials and MEUUs. By contrast, the system of interaction between the state and non-SOEs and households/individuals relies significantly on market-based mechanisms, financial incentives, lifestyle orientation, and information dissemination, although juridical and administrative powers also play a role via the setting of EC standards and various kinds of quota. Chapter 4 reveals that the centre–localities/SOEs interface is by far the most pivotal relationship in terms of carbon government precisely because of the effectiveness of the ECERTRS. The latter depends heavily on not only the planning system prioritizing MLCT, but also regularized and regulated MA&E mechanisms and strong sanctions.

Then, in Chapters 5 and 6, we investigated the practical effects of China's carbon governmentality in two localized systems: the financial system and cities. In Chapter 5, we first examined how China has financed its massive investment programme into EC and RE technologies in the past decade. Our analysis suggests that this has been accomplished mainly by affecting the behaviours of FIs and SNGs, rather than through direct allocation of public resources, although there have been significant (around 10 per cent) inputs from public finance. What has been most impactful is the Chinese authorities' concerted efforts in greening the entire financial system through a complex green finance governmentality. China has actively promoted green credits since 2007 and green bonds since 2016. As a result, the green credit balance reached RMB 11.95 trillion (USD 1.93 trillion) by the end of 2020, whereas green bonds raised more than a trillion yuan between 2016 and 2019. Our estimate is that budgetary allocation, the RE Levy, green credits and green bonds financed at least 62 per cent of all MLCT investment over 2012–19, with green credits alone contributing more than 50 per cent. The fact that green credits have accounted for 6–7 per cent of all loans from 2013 onwards is an indication of the significant transformation of the financial system that has taken place. It is likely though that greening finance has expanded the hold of the state and strengthened the position of the SOEs, as the latter have been able to take better advantage of green finance for structural reasons such as their dominance among FIs and higher credit rating.

Chapter 6 investigated the effects of carbon governmentality in three Chinese cities (Shanghai, Qingdao and Hangzhou), all aiming to peak their CO_2 emissions by 2020, ten years ahead of the rest of the country. This inquiry draws significantly on information collected from my own fieldwork. The principal findings are as follows. First, these cities have achieved faster paces of decarbonization than centrally required (see Table 6.3). Second, their performance is explained to a large extent by their positive response to the central carbon governmentality and the development of carbon governmentalities of their own. The latter are oriented towards meeting the centrally disaggregated and allocated targets and coordinating a smooth LCT in the local context. On the one hand, the cities have endeavoured to localize the centre's political rationalities on MLCT and sought to align their pre-existing local agendas with the MLCT agenda, with varying degrees of success. On the other hand, the city authorities have developed their own GTTs to promote MLCT, targeting mainly MEUUs, public institutions, government officials, households, and, increasingly, individuals. It is evident that the city authorities have endeavoured to achieve their binding planning targets and comply with the operative requirement of the ECERTRS; they have also invested heavily in institution building, GHG inventory compilation, strategic planning, low-carbon infrastructural investment, industrial policy, and international collaboration. Third, the MLCT process has been associated with a steady slow-down of economic growth across the three cities during the past decade. This at least partially resulted from the shrinkage of the energy-intensive, high-value-adding manufacturing activities, which have borne the brunt of industrial restructuring under the state's industrial policy geared towards MLCT. Fourth, as the experience of Ant Forest shows, the Chinese society, especially among young urban consumers, has started to develop a green and low-carbon ethics. Fifth and finally, these cities are deploying large-scale investment, in the order of 2–5 per cent of GDP, or 3–10 per cent of fixed asset investment (FAI), for the purpose of MLCT. Much of this investment is financed through green credits and green bonds, which suggests that the centrally developed green finance governmentality has had its expected effect.

It can be concluded that the introduction of the binding planning targets for carbon intensity reduction and RE development for the first time in the 12th FYP (2011–15), and the reinforcement of such targets through the 'double control' mechanism and an integrated approach to climate actions during the 13th FYP (2016–20) period have ushered in a decade of accelerated MLCT in China. What should be stressed is that it is the articulation between strong political rationalities and the wide-ranging carbon GTTs that lies at the heart of this performance. ECERTRS, the linchpin of China's carbon governmentalities, relies on the mainstreaming of MLCT objectives in various kinds of plans, in the CES, and a rigorous MA&E system. Carbon governmentalities have capitalized on the development of a whole complex of discourse and knowledge about MLCT, including new performance criteria, emissions inventories, scenario analysis, MLCT roadmaps, international standards, and the like. The defining features of these governmentalities include their strategic alignment with the goal of consolidating socialism through the pursuit of a more sustainable development model including MLCT, the combined deployment of juridical, administrative and governmental powers, the simultaneous use of planning and market mechanisms, and the ECERTRS as a technology of performance. Key elements of these carbon governmentalities are summarized in Table 7.1.

It is also clear that the turning point of 2011 coincides with the onset of China's 12th FYP, which introduced for the first time binding planning targets for carbon intensity reduction and the development of RE, the assumption of the top leadership position by Xi Jinping in 2012, and the launch, also in 2012, of the 'Integrated Reform Plan for Promoting Ecological Progress', which seeks to institutionalize this concept introduced initially by President Hu in 2007. These three factors reinforced each other thanks to President Xi's long-standing support for green development.

The last six chapters also show that China's MLCT process in the 2010s was not smooth sailing and that it required close monitoring and continuous steering and, sometimes, rowing from state authorities. At various points and in numerous fields, it suffered from slow starts, setbacks, fluctuations, and shortfalls. However, the political authorities have been willing to address them. More crucially, they have become adept at bringing in new discourse and GTTs to deal with problems as they arise. To illustrate this, it suffices to cite three examples from the chapters. First, the introduction of binding planning targets in the 11th FYP did not have its expected effect until the ECERTRS was put in place in 2007. Second, the government's EV programme suffered from a slow start. It only gained impetus when private users became subsidy beneficiaries in 2012, a market-assisted allocation system of vehicle licences made the free licenses available to EV purchasers valuable, and when EV manufacturers gained privileged access to green credits from 2013. Third, the problem of high RE curtailment was turned around by a cocktail of regional RE quotas, conditionality for new RE investments, RE consumption's exemption from the 'double control' mechanism, and strengthened monitoring and inclusion into ECERTRS.

The chapters also suggest that, although China's transition towards a low-carbon future is secure, it will be hard to sustain its current pace without further breakthroughs. A three-pronged deceleration mechanism has emerged during the 13th FYP period (2016–20), comprising slow-down of economic growth, its negative repercussions on the continuing reduction of carbon intensity through investment in low-carbon technologies, and regional protectionism. First, while slower growth should curtail the growth of total emissions, it is also making it harder for China to meet its carbon intensity reduction target. Chandler and Wang (2009) warned earlier that, to achieve an annual rate of decarbonization of 4 per cent, the carbon/GDP elasticity only needs to fall to 0.5 if GDP growth is 8 per cent per annum, which is achievable; when the GDP growth is only 6 per cent, however, the elasticity will have to fall to 0.33 to achieve the same rate of decarbonization. China's 1 per cent decarbonization in 2020 seems to confirm this, although the over-fulfilment in the first four years of the 13th FYP may also be a contributory factor. Second, slower growth has also resulted in stagnant RE Levy revenue, which in turn

Table 7.1   *Key elements of carbon governmentalities in China*

Analytical category	Key questions	Key elements
Political rationalities		
Field of visibility	What is illuminated?	Sources of emission by industry, region and enterprise
		Relevant ministries and agencies
		Eight principal subject groups (especially SNGs, MEUUs, and FIs)
		SOEs
	Problems to be solved	High carbon intensity of the economy
	Objectives to be sought	To meet China's international pledges on mitigation, while continuing to develop the economy and realize the long-term national goals of socialist modernization and local prosperity
Forms of knowledge	Which forms of thought arise from and inform the activity of governing?	Modernization and industrialization
		The Scientific Outlook on Development (SOD), including sustainable development
		Ecological Civilization (i.e., building a resource-efficient and environment-friendly economy)
		Nature as resources and assets
		Planning and rationalization
		Political economy (efficiency; competitiveness; performance-based legitimacy)
		Sciences of climate and energy systems
		Techno-economic paradigm
Formation of identities and enhancement of agency	What forms of self and persons are presupposed by practices of government?	Calculative cadres (primary)
		Rational producers and consumers (secondary)
		Rational households and individuals (tertiary)
	Which transformations are sought?	Responsible carbon managers

	Analytical category	Key questions	Key elements
Governing techniques and technologies	Technical aspects	By what instruments, procedures and technologies is rule accomplished?	FYPs/sectoral plans/dedicated work plans/long-term strategies
			Law: Energy Conservation Law; Renewable Energy Law
			Institutional restructuring and capacity building (including changes to the performance evaluation system for top officials and SOE managers)
			Technologies of agency: empowering and enabling especially SNGs, MEUUs and consumers
			Technologies of performance: ECERTRS for SNGs and ECTRS for MEUUs
			Compilation of GHG inventories by regions (down to urban districts in selected provinces)
			Industrial policy: Phasing-out of 'two highs and one surplus' industries and fostering of low-emission industries/technologies/products
			Regulation: mandatory energy efficiency standards; energy and climate statistics; energy use and EC audit; FAI censorship
			RE quotas for provinces and energy suppliers; 'double-control' mechanisms for SNGs
			RE Levy and multi-level special state funds for MLCT
			State investment in low-carbon technologies and subsidies for RE and EC
			Market mechanisms: feed-in tariffs; energy pricing reform; EPC; carbon emissions trading
			Greening of FIs and financial markets, products, and instruments (including incentives for green credits and green bonds)
			Experimental piloting of many kinds in different sectors and regions (including EUA trading and auction and lottery allocation of vehicle licenses)
			Low-carbon development strategies, scenarios, and roadmaps based on GHG inventories and cost-benefit analysis of mitigation options
			Mass mobilization exploring aesthetics, ethics, and ICT

led to diminishing RE subsidies from the central government from 2016. As a matter of fact, slow-down in the growth of the overall budgetary revenue is mirrored in the decline in subsidies and public finance allocations for EC and RE (see Table 5.1). Third and finally, while the current system relies heavily on 'management by region', it is prone to excessive inter-region competition and regional protectionism. It has aggravated problems of diminishing subsidies due to a lack of efforts of collecting RE Levy revenues (Everbright Securities 2019) and experienced high curtailment of RE capacities as regional governments tried to protect/maintain their own RE capacities when demand for energy slowed down (Li et al. 2017). So far it has hindered the efficient exploitation of RE resources and led to occasionally wasteful inter-region competition in MLCT-related manufacturing activities.

A combined effect of these factors is a likely slow-down in the pace of MLCT in coming decades unless additional GTTs are put in place and new revolutionary technologies emerge. Whether the national ETS will prove to be one of these breakthroughs is impossible to predict at this point. In addition, the governing objects are also evolving. While decarbonizing industry has made a large contribution to the performance so far, the potential for future decarbonization will have to be sought increasingly in transport and buildings, where the scope for mitigation may be not as great due to the lock-in effects of the existing sprawling urban forms and infrastructure.

This combination means that additional efforts need to be made to explore new kinds of GTTs. A comparison of the general GTTs as described in Chapter 4 and those in Chapter 5 as applied to the financial system is instructive. It appears that the government has adopted different tactics for the tasks of ensuring binding MLCT planning targets on the one hand and greening the financial system on the other. While the ECERTRS is dominant in the former, a set of softer GTTs has been used for the latter. By establishing the GCSS and GCSSS, setting green finance standards and taxonomies, conducting green bank self-assessment and external evaluation, incorporating green credits and bonds into the MLF and MPA, and encouraging information disclosure by FIs, Chinese financial regulators have made the performance of FIs, mainly banks, calculable, measurable, comparable, and rewardable. It is also clear though that the latter is not as effective as the former. A lack of asymmetric growth by green credits and the tiny proportion of green bonds in the bond market are testaments to this.

EVALUATING THE THEORETICAL IMPLICATIONS

Let us now reflect on the theoretical implications of the above findings. One question to address is whether these shed new light on aspects of the Chinese experience that were previously underexplored or unexplained. To start with, it may be useful to contrast our findings with what Dryzek (2013) has said about China's approach to the environment because both studies attend to discourses and their links with characteristic policy instruments and institutions. He characterizes China as a case of administrative rationalism, where top-down planning that sets targets centrally also specifies the means to achieve them (Dryzek 2013, p. 83). He attributes this kind of environmental planning in China to a lack of the enforcement of environmental laws and regulations in the country. Our analysis in this book contradicts his characterization by showing that, by the late 2000s, the Chinese MLCT government system has become much more complex than that on at least three levels. First, there is now much greater emphasis on the enforcement of environmental laws and regulations. Moreover, these are combined with the pre-existing cadre evaluation system under which leading cadres are held legally responsible for meeting binding planning targets related to MLCT. Indeed, the laws on EC and RE have played a key role in introducing the feed-in tariff (FiT) scheme, improving energy efficiency and supporting the development of renewables. Second, the commitment towards the SOD (2007) and the institutionalization of Ecological Civilization (2012) (above and beyond Ecological Modernization) show that the Chinese system has moved beyond mere problem-solving to the imaginative pursuit of sustainability. Third, as Chapters 4 and 6 show, the central carbon governmentality in China provides significant scope for local autonomy and initiatives on the one hand and for the deployment of market mechanisms on the other hand. While the former is made possible by the tradition of RDA, the latter draws its legitimacy from the conception of SME, introduced in the early 1990s.

It is also instructive to compare our findings with existing wisdoms on China's MLCT performance, especially those highlighted in Chapter 1. First, this study shows that China's progress goes well beyond the field of clean energy. To a large extent, the success of China hinges more on its effectiveness in improving energy efficiency than on the development of RE partly because of curtailment problems that peaked around 2016. The endeavour on EC dating back to the 1980s has become institutionalized through the coupling of FYPs and the ECERTRS. On the other hand, contrary to the assessments of Ploberger (2013) and Andrews-Speed (2012), China has demonstrated signif-icant adaptive capacities in response to the MLCT challenge. Its continuous strengthening of policy goals and measures over a 15-year period (see Table 1.1 and Table 4.2) and its pathbreaking approach to the greening of the finan-cial system, as discussed in Chapter 5, testify to this adaptability. Moreover, contradicting Engels' (2018) claim, such success is evidently based on an increasingly integrative strategic switch from around 2014. Our findings also challenge previous suggestions that whatever China has been able to achieve can be explained by how much economic, social, and political costs have been spent (Andrews-Speed and Zhang 2019; Toke 2017). This book shows that China's enhanced performance since the early 2010s can be better explained by the development of its carbon governmentalities than the costs incurred. In financial terms, MLCT has claimed around 6–7 per cent of all bank loans and less than 1 per cent of all the capital raised from bond issuance during the decade. Moreover, as Talcott Parsons' (1963) framework of political power has suggested, China's MLCT efforts may have added to its power and political capital, just like commercial bankers do through credit creation. China's rising profile on climate actions and green finance internationally illustrate that the Chinese leadership's foresight in committing to MLCT and green finance has paid off.

Our analysis further suggests that China is transitioning from a developmental state into a carbon-governmental state. While the former prioritizes economic development, in parallel with Meadowcroft's (2005) definition of the ecostate, the latter may be defined as a state that mainstreams carbon control into all its work; it is characterized by the harmonization of its development goal with carbon control, the carbonization of its legislative, bureaucratic and planning systems, and the development of a fully fledged governing system comprising technologies of performance for public authorities/officials and technologies of agencies for ECUs, MEUUs, households and individuals. In this connection, it is useful to compare what happens under China's carbon governmentalities (see Table 7.1) with the so-called 'advanced liberal government' (ALG), prevalent in the developed Western countries (Oels 2005). While ALG targets individuals and social groups, guards against 'excessive' state bureaucracy, governs by using markets as an organizing principle and presupposes calculating individual entrepreneurs, Chinese carbon government targets many groups, but most importantly SNGs, government officials, MEUUs, and FIs, uses a mix of CAC, planning, market mechanisms, and ethics, and presupposes calculating officials, producers, and consumers. In other words, the latter creates a much wider field of visibility and uses a far wider range of GTTs. Perhaps most crucially, China's carbon governmentalities place SNGs and government officials at the forefront of decarbonization.

It is necessary to point out that China's success in MLCT so far is not a testimony to the efficacy of its authoritarianism, defined as unrestrained state domination, but the state's ability to make commitments, to operationalize them and to mobilize different types of resources and tools to ensure attainment of goals. It involves leadership as defined by Parsons (1963). What Parsons' analysis lacks in terms of the mechanisms of determination of system goals (Savage 1978) or the social processes whereby the value-consensus is maintained (Giddens 1968), the analytics of government illuminates. The latter shows that, in China, the goal consensus is reached and maintained through continuous reference to the overall goal, namely the 'Three Criteria' and 'Three Represents' that the founding fathers of China's reform programme have identified (see Chapter 3). However, a governmentality perspective offers more than a highlight on political leadership. It demonstrates strongly how goals are rendered achievable through the development of political rationalities and the concerted deployment of GTTs.

LESSONS: THE ROLE OF CARBON GOVERNMENTALITY IN MLCT

Now it is time for us to address the third question, namely, what other nation states could learn from the Chinese experience. Despite its unique characteristics and circumstances, China has several lessons to offer those nation states poised to embark on a journey of fast decarbonization. These include most importantly the value of political leadership, the importance of accentuating the state's positive power by creating the right kinds of discourse, knowledge and GTTs for MLCT, the need to balance MLCT with economic development, and to make full use of all kinds of power.

First, the case of China demonstrates that political ambition and leadership can pay off. Here, as Parsons (1963) has argued, the conception of non-zero-sum politics is a potentially useful model to adopt in MLCT. Just like commercial banks can create more value through credit creation, if successful, political leaders daring to aim for higher goals and making difficult commitments can reap fruits for both themselves and the society at large. Leadership is an integral part of the positive power that Foucault emphasizes. However, to be successful, political leaders must use all the tools available to them, be they coercion, inducement, influence, or moral commitments.

Second, China's experience shows that there is a lot that the state can do for MLCT and that both political rationalities and GTTs matter in addressing climate change. Discourse and knowledge have certainly played an important role. From its initial inattention to climate change, to the abandonment of its focus on defending its emission rights in the first half of the 2000s, and to its eventual conceptualization of climate-related work as a lever for building an Ecological Civilization, the change in mindset and discourse has provided a strong underpinning for China's increased efforts and effectiveness in MLCT in the 2010s. However, China's experience shows even more strongly that GTTs are crucial to policy effectiveness. Without the persistent and widespread use of planning, the implementation of industrial policy favouring service industries and discouraging resource-heavy and emission-intensive manufacturing activities, the restructuring of the bureaucracy, the re-definition of performance criteria for SNGs/SOEs officials, the spotlight on MEUUs and FIs, incentives and inducements for producers, households and individuals to save energy and adopt climate friendly products and technologies, and the integration of reformed performance indicators, MA&E, and sanction through the ECERTRS, it would hardly have been possible to achieve its various policy objectives. A crucial element in this process is constant monitoring of the implementation of planning targets. In several instances, annual monitoring led the state to discover shortfalls and introduce additional plans or measures to bring a straying course back on track.

Third, China's numerous setbacks during the past decade, including the slow-down in its economic growth, diminishing investment for RE and high RE curtailment, show that the relationship between MLCT and economic development is fraught with difficulty. It requires constant attention to the question of 'how', that is, how to make this relationship work. China's relative success in coupling MLCT with economic development so far is largely thanks to its continuous emphasis on modernization and productivity during MLCT and its efforts to exploit new investment and growth opportunities that have emerged during MLCT. Yet, the opportunities of doing so may be diminishing as the 'low-hanging fruits' run out.

Fourth, the case of China demonstrates the significant and often underestimated potential of CAC mechanisms and administrative power, when they are deployed alongside juridical and governmental powers. While China has shown remarkable keenness in experimenting with market and other governmental mechanisms, it is in the use of CAC mechanisms that it has excelled so far. However, it would be a mistake to consider the Chinese regime of practices merely as a CAC governance system. For example, rather than relying on technical fixes, as Chapter 3 demonstrates, China has put a great deal of emphasis on fostering green and low-carbon rationalities. On the other hand, it has managed to avoid panaceas and suppression of local initiatives by encouraging local experimental pilots, market-oriented experimentations, and continuous policy learning. Ultimately its performance can be attributed to the emergence and development of carbon governmentalities at both central and local levels and in different areas. The key mechanism here is the centrally set MLCT targets compelling local authorities and other stakeholders to look for ways of conducting and financing MLCT that fit local conditions. Moreover, as demonstrated by the experience of Ant Forest, the discourses motivated corporations and private individuals to join the search for MLCT solutions by exploring the ethic of the aesthetic. Just as Dean (2002) has argued that liberal rationalities and authoritarian measures are far from incompatible, the Chinese experience shows that authoritarian rationalities can be compatible with liberal measures.

In addition, the experience of China shows that there is a great deal of scope for learning and for deploying and creating knowledge about measuring, monitoring, and regulating carbon behaviour in support of MLCT. In this regard, as shown in Chapter 6, the roles of sub-national authorities and international partners are especially important. It is suggested that we are witnessing the emergence of another kind of power – carbon power. While biopower is built on the knowledge about the population – its growth, health, education, and the like – carbon power can be similarly defined as a form of power that is built upon the knowledge about MLCT, including how to control carbon emissions, whom to rely on and how to affect their behaviours through the furtherance and application of sciences and discourses.

Ultimately, the most important lesson from China and from this book is that carbon governmentality and the carbon governmental state really matter for decarbonization. Yet there is so much more we need to learn about them...

References

21st CBH (2016), 'Third-Batch Low-Carbon City Pilots to Be Announced in November', accessed 11 April 2020 at http://epaper.21jingji.com/html/2016-10/27/content_49161.htm.

Aizawa, M. and C. Yang (2010), 'Green Credit, Green Stimulus, Green Revolution? China's Mobilization of Banks for Environmental Cleanup', *The Journal of Environment & Development*, 19 (2), 119–44.

An, G. J. (2021), 'Green Finance Innovation Paths under Carbon Neutral Goal', accessed 8 March 2021 at http://www.tanpaifang.com/tanjinrong/2021/0306/76956_4.html.

Andrews-Speed, P. (2012), *The Governance of Energy in China*, Basingstoke: Palgrave Macmillan.

Andrews-Speed, P. and S. Zhang (2019), *China as a Global Clean Energy Champion: Lifting the Veil*, Singapore: Palgrave Macmillan.

Asian Development Bank (ADB) (2018), *50 Climate Solutions from Cities in the People's Republic of China: Best Practices from Cities Taking Action on Climate Change*, Manila: ADB. Downloadable from: https://www.adb.org/publications/50-climate-solutions-prc-cities.

Avelino, F. (2017), 'Power in Sustainability Transitions: Analysing Power and (Dis) empowerment in Transformative Change Towards Sustainability', *Environmental Policy and Governance*, 27 (6), 505–20.

Avelino, F. and J. Rotmans (2009), 'Power in Transition: An Interdisciplinary Framework to Study Power in Relation to Structural Change', *European Journal of Social Theory*, 12 (4), 543–69.

Avelino, F. and J. Rotmans (2011), 'A Dynamic Conceptualization of Power for Sustainability Research', *Journal of Cleaner Production*, 19 (8), 796–804.

Baidu (2021a), Qingdao Energy Conservation Supervision Centre, accessed 9 March 2021 at https://baike.baidu.com/item/青岛市节能监察中心.

Baidu (2021b), Ten Cities Thousand Vehicles Project, accessed 15 June 2020 at https://baike.baidu.com/item/十城千辆工程/451317?fr=aladdin.

Bank of Qingdao (2019), *Special Report on the Use of Green Bond Proceeds* (2019 Q3), accessed 3 May 2020 at https://www.chinabond.com.cn/cb/cn/ywcz/fxyfxdf/zqzl/syyhz/ptz/qtggtz/20191025/152985439.shtml.

Barry, J. and R. Eckersley (eds.) (2005), *The State and the Global Ecological Crisis*, Cambridge, MA: MIT Press.

Bevir, M. (2012), *Governance: A Very Short Introduction*, Oxford: Oxford University Press.

Bray, D. (2005), *Social Space and Governance in Urban China: The Danwei System from Origins to Reform*, Stanford, CA: Stanford University Press.

Brødsgaard, K. E. and K. Rutten (2017), *From Accelerated Accumulation to Socialist Market Economy in China: Economic Discourse and Development from 1953 to the Present*, Leiden: Brill.

Bulkeley, H. (2010), 'Cities and the Governing of Climate Change', *Annual Review of Environment and Resources*, 35 (1), 229–53.

Bulkeley, H. A., V. C. Broto and G. A. S. Edwards (2014), *An Urban Politics of Climate Change: Experimentation and the Governing of Socio-Technical Transitions*, London: Routledge.

Bullard, N. (2019), 'China Is Winning the Race to Dominate Electric Cars', accessed 17 July 2020 at https://www.bloomberg.com/opinion/articles/2019-09-20/electric-vehicle-market-so-far-belongs-to-china.

C40 Cities (2017), 'Qingdao – Optimizing the Energy Use Structure of Heating to Achieve the Goal of Clean Heating', accessed 10 March 2020 at https://www.c40.org/awards/2017-awards/profiles/135.

C40 and Lawrence Berkeley National Laboratory (LBNL) (2018), *Constructing a New, Low-Carbon Future: How Chinese Cities Are Scaling Ambitious Building Energy-Efficiency Solutions*. China Buildings Programme Launch Report, accessed 11 May 2020 at https://www.c40.org/researches/constructing-a-new-low-carbon-future-china.

Castán Broto, V. (2017), 'Urban Governance and the Politics of Climate change', *World Development*, 93, 1–15.

Cerutti, E. and M. Obstfeld (2018), 'China's Bond Market and Global Financial Markets', *IMF Working Papers*, 18 (253).

Chandler, W. and Y. C. Wang (2009), 'Memo to Copenhagen: Commentary is Misinformed – China's Commitment is Significant', accessed 17 May 2020 at https://carnegieendowment.org/files/Memo_to_Copenhagen_edits_Revised_12-14-091.pdf.

Chen, T.-J. (2016), 'The Development of China's Solar Photovoltaic Industry: Why Industrial Policy Failed', *Cambridge Journal of Economics*, 40 (3), 755–74.

Chen, Z.-Y. and N. Jiang (2020), 'A Review of China's Green Bond Market 2019', accessed 29 February 2020 at http://greenfinance.xinhua08.com/a/20200110/1907055.shtml?f=areco.

Cheng, J. H. and J. J. Guo (2016), 'Hangzhou's "Six-in-One" Low-Carbon Urban Development Practices and Lessons', in Green Low-Carbon Development Think-Tank Partnership (GLCDTTP) (ed.), *Policies and Practices in China's Low-Carbon Pilot Cities: Bring the Emissions Peak Forward*, Beijing: Science Press, pp. 206–18.

Chien, S. S. (2013), 'Chinese Eco-Cities: A Perspective of Land-Speculation-Oriented Local Entrepreneurialism', *China Information*, 27 (2), 173–96.

China Banking Association (CBA) (2018), 'Circular on Implementation Plan of Green Bank Evaluation in the Banking Industry (Trial)' (No. 171), accessed 25 April 2020 at http://www.hqwx.com/web_news/html/2018-3/15208259789600.html.

China Banking Regulatory Commission (CBRC) (2007), 'Guidance on Lending for Energy Conservation and Emissions Reduction' (No. 83), accessed 25 August 2020 at http://www.gov.cn/zwhd/2007-12/29/content_846830.htm.

China Banking Regulatory Commission (CBRC) (2012), 'Circular on Issuing Green Credit Guidelines' (No. 4), accessed 26 April 2020 at http://www.lawinfochina.com/display.aspx?lib=law&id=9239&CGid=.

China Banking Regulatory Commission (CBRC) (2013), 'Green Credit Statistical System' (No. 185), accessed 26 April 2020 at http://www.cbrc.gov.cn/chinese/home/docView/F0E89A3240984465BFEF1E3D01316D5B.html.

China Banking Regulatory Commission (CBRC) (2014), 'Key Performance Indicators of Green Credit Implementation' (No. 186), accessed 26 April 2020 at http://zfs.mee.gov.cn/hjjj/gjfbdjjzcx/lsxdzc/201507/t20150716_306812_wap.shtml.

China Banking Regulatory Commission (CBRC) (2018a), 'Green Credit Data of 21 Major Domestic Banks from 2013 to June 2017', accessed 2 March 2021 at http://www.cbirc.gov.cn/cn/view/pages/ItemDetail.html?docId=171047.

China Banking Regulatory Commission (CBRC) (2018b), *2017 Annual Report: Annex 4*. Downloadable from: http://www.cbirc.gov.cn/cn/view/pages/ItemDetail_gdsj.html?docId=24216&docType=0.

China Center for Modernization Research (CCMR) of Chinese Academy of Sciences (CAS) (2007), *China Modernization Report 2007: Study on Ecological Modernization*, Beijing: Beijing University Press.

China Bond Rating Ltd (2018), *An Inquiry into the Mechanisms and Practices of Green Finance Supporting Industrial Structure Upgrading*, Special Report No. 65 of 2018 (No. 603 of All Issues), accessed 27 April 2020 at http://www.twoeggz.com/news/10264321.html.

China Financial and Economic News (2019), 'US $100 Million Green Climate Fund Loan to Support ADB's Shandong Climate Project', accessed 13 May 2020 at http://www.cfen.com.cn/dzb/dzb/page_3/201911/t20191121_3426552.html.

China Securities Regulatory Commission (CSRC) (2017), 'Guidance of the CSRC on Supporting the Development of Green Bonds', accessed 16 April 2020 at http://www.csrc.gov.cn/pub/zjhpublic/G00306201/201703/t20170303_313012.htm.

Chinese Communist Party Central Committee (CCPCC) (2013), 'Decision on Several Major Issues Concerning Comprehensively Deepening Reform', accessed 23 April 2020 at http://www.ce.cn/xwzx/gnsz/szyw/201311/18/t20131118_1767104.shtml.

Chinese Communist Party's Central Committee (CCPCC) and the State Council (2015a), 'Integrated Reform Plan for Promoting Ecological Progress', accessed 24 January 2021 at http://www.gov.cn/gongbao/content/2015/content_2941157.htm.

Chinese Communist Party's Central Committee (CCPCC) and the State Council (2015b), 'Suggestions of the CCP Central Committee and the State Council on Accelerating the Construction of Ecological Progress', accessed 29 March 2020 at http://www.gov.cn/xinwen/2015-05/05/content_2857363.htm.

Chinese Communist Party Central Committee General Affairs Office (CCPCCGAO) and State Council General Affairs Office (SCGAO) (2015), 'Measures for Investigating the Responsibility of Leading Party and Government Cadres for Eco-Environmental Damage (Trial Implementation)', accessed 2 March 2021 at http://www.xinhuanet.com/politics/2015-08/17/c_1116282540.htm.

Chinese Communist Party Central Committee General Affairs Office (CCPCCGAO) and State Council General Affairs Office (SCGAO) (2016), 'Assessment Methods of Ecological Progress Construction Objectives', accessed 24 January 2021 at http://www.xinhuanet.com/politics/2016-12/22/c_1120169808.htm.

Chinese Communist Party Central Committee Organization Department (CCPCCOD) (2009), 'Interim Procedures for Comprehensive Assessment and Evaluation of Subnational Party and Government Leading Corps and Cadres' (No. 13), accessed 11 September 2020 at http://wsqdjw.gov.cn/zcfg/201211/t20121121_745004.html.

Christensen, J. and A. Olhoff (2019), 'Lessons from a Decade of Emissions Gap Assessments', accessed 17 May 2020 at https://wedocs.unep.org/bitstream/handle/20.500.11822/30022/EGR10.pdf?sequence=1&isAllowed=y.

CHYXX (2018), 'Development Trend Analysis of China's New Energy Vehicle Industry in the Next 3–5 Years', accessed 8 March 2021 at http://www.chyxx.com/industry/201804/632950.html.

Climate Action Tracker (CAT) (2021), 'Comparison of NDC Targets', accessed 7 April 2021 at https://climateactiontracker.org/climate-target-update-tracker/china/.

Climate Bonds Initiative (CBI) (2019a), 'China Green Bond Market Newsletter H1 2019', accessed 28 April 2020 at https://www.climatebonds.net/resources/reports/china-green-bond-market-newsletter-h1-2019.

Climate Bonds Initiative (CBI) (2019b), 'Green Bonds: The State of the Market 2018', accessed 26 April 2020 at https://www.climatebonds.net/resources/reports/green-bonds-state-market-2018.

Climate Bonds Initiative (CBI) (2021), 'Sustainable Debt: Global State of the Market 2020', London. Downloadable from: https://www.climatebonds.net/resources/reports/2021.

Climate Bonds Initiative (CBI) and China Central Depository & Clearing Co. Ltd (CCDC) (2019), 'China Green Bond Market 2018', accessed 28 April 2020 at https://www.climatebonds.net/files/reports/china-sotm_cbi_ccdc_final_en260219.pdf.

Climate Bonds Initiative (CBI) and China Central Depository & Clearing Research Centre (CCDC Research) (2020), *China Green Bond Market 2019 Research Report*, London.

China National Radio (CNR) (2016), 'The 13th Five Year Plan: Establishing and Improving the Initial Distribution System of Energy Use Allowances, accessed 28 August 2020 at http://china.cnr.cn/ygxw/20160709/t20160709_522629048.shtml.

Cox, M. (2016), 'The Pathology of Command and Control: A Formal Synthesis', *Ecology and Society*, 21 (3), 33.

Cruikshank, B. (1993), 'Revolutions Within: Self-Government and Self-Esteem', *Economy and Society*, 22 (3), 327–44.

Cunningham, E., T. Saich and J. Turiel (2020), 'Understanding CCP Resilience: Surveying Chinese Public Opinion Through Time', accessed 4 August 2020 at https://ash.harvard.edu/publications/understanding-ccp-resilience-surveying-chinese-public-opinion-through-time.

Dean, M. (1996), 'Putting the Technological into Government', *History of the Human Sciences*, 9 (3), 47–68.

Dean, M. (1999), *Governmentality: Power and Rule in Modern Society*, London: Sage Publications.

Dean, M. (2002), 'Liberal Government and Authoritarianism', *Economy and Society*, 31 (1), 37–61.

Dean, M. (2010), *Governmentality: Power and Rule in Modern Society*, 2nd edition, London: Sage Publications.

Deng, X. (1993), *Selected Works of Deng Xiaoping: Vol. 3*, Beijing: People's Publishing House.

Deng, Y. (2020), 'Numbers of New Energy Passenger Cars Licensed in May', accessed 24 June 2020 at https://auto.gasgoo.com/news/202006/24I70189926C108.shtml.

Department of Trade and Industry (DTI) (2003), *Our Energy Future: Creating a Low Carbon Economy*, Norwich: The Stationery Office.

Dirlik, A. and X. Zhang (1997), 'Introduction: Postmodernism and China', *Boundary 2*, 24 (3), 1–18.

DRCNET (2020), 'Bank Green Credit Outstanding Balance (End of 2018)', accessed 21 February 2020 at http://greene.drcnet.com.cn/web/.

Dryzek, J. (2013), *The Politics of the Earth: Environmental Discourses*, Oxford: Oxford University Press.

Duan, M., T. Tian, Y. Zhao and M. Li (2018), 'Interactions and Coordination between Carbon Emissions Trading and Other Direct Carbon Mitigation Policies in China', in Y. Qi and X. Zhang (eds.), *Annual Review of Low-Carbon Development in China*, Beijing: Social Sciences Academic Press (China).

Economic Information Daily (EID) (2016), '100 Billion Green Financial Bonds Approved', accessed 21 February 2020 at http://bond.jrj.com.cn/2016/01/25005020471905.shtml.

Edin, M. (2003), 'State Capacity and Local Agent Control in China: CCP Cadre Management from a Township Perspective', *The China Quarterly*, 173, 35–52.

Elazar, D. (1995), 'From Statism to Federalism: A Paradigm Shift', *Publius: The Journal of Federalism*, 25 (2), 5–18.

Elliott, C. and L.-Y. Zhang (2019), 'Diffusion and Innovation for Transition: Transnational Governance in China's Green Bond Market Development', *Journal of Environmental Policy & Planning*, 21 (4), 391–406.

Energy Foundation China (EFC) (2019), 'Energy Foundation China: 20 Years, 10 Stories', Beijing. Downloadable from: http://www.efchina.org.

Engels, A. (2018), 'Understanding how China is Championing Climate Change Mitigation', *Palgrave Communications*, 4 (1), 101.

Escalante, D., J. Choi, N. Chin, Y. Cui, and M. L. Larsen (2020), 'The State and Effectiveness of the Green Bond Market in China'. *Climate Policy Initiative*. Downloadable from: https://www.climatepolicyinitiative.org/publication/green-bonds-in-china-the-state-and-effectiveness-of-the-market/.

Ettlinger, N. (2011), 'Governmentality as Epistemology', *Annals of the Association of American Geographers*, 101 (3), 537–60.

Everbright Securities (2019), *The Renewable Energy Development Special Fund: Its History and Present*, accessed 7 April 2020 at https://www.vzkoo.com/news/1830.html

Feindt, P. H. and A. Oels (2005), 'Does Discourse Matter? Discourse Analysis in Environmental Policy Making', *Journal of Environmental Policy & Planning*, 7 (3), 161–73.

Forward Business and Intelligence Limited (FBIL) (2021), 'An Analysis of China's Green Credit Market 2020: Current State and Structure', accessed 17 February 2021 at https://www.sohu.com/a/450602767_120868906.

Foucault, M. (1970), 'The Order of Discourse'. Reprinted in R. Young (ed.), *Untying the Text: A Post-Structuralist Reader*. London: Routledge, 1983.

Foucault, M. (1977), *Discipline and Punish: The Birth of the Prison*, trans. A. Sheridan, New York: Vintage Books.

Foucault, M. (1978), *The History of Sexuality Volume I: An Introduction*, trans. R. Hurley, New York: Pantheon Books.

Foucault, M. (1982), 'The Subject and Power', *Critical Inquiry*, 8 (4), 777–95.

Foucault, M. (1984), 'Truth and Power', in P. Rabinow (ed.), *The Foucault Reader: An Introduction to Foucault's Thought*, London: Penguin Books, pp. 51–75.

Foucault, M. (1991a), 'Governmentality', in G. Burchell, C. Gordon and P. Miller (eds.), *The Foucault Effect: Studies in Governmentality, with Two Lectures by and an Interview with Michel Foucault*, London: Harvest Wheatsheaf, pp. 87–104.

Foucault, M. (1991b), 'Questions of Method', in G. Burchell, C. Gordon and P. Miller (eds.), *The Foucault Effect: Studies in Governmentality, with Two Lectures by and an Interview with Michel Foucault*, London: Harvest Wheatsheaf, pp. 73–86.

Foucault, M. (1991c), 'Politics and the Study of Discourse', in G. Burchell, C. Gordon and P. Miller (eds.), *The Foucault Effect: Studies in Governmentality, with Two Lectures by and an Interview with Michel Foucault*, London: Harvest Wheatsheaf, pp. 53–72.

Foucault, M. (2008), *The Birth of Biopolitics: Lectures at the Collège de France, 1978–1979*, Basingstoke: Palgrave Macmillan.

Friedlingstein, P. et al. (2020), The Global Carbon Budget 2020, Earth System Science Data. Available athttps://essd.copernicus.org/articles/12/3269/2020/.

FS-UNEP (2019), *Global Trends in Renewable Energy Investment 2019*.

FS-UNEP (2020), *Global Trends in Renewable Energy Investment 2020*.

Gao, J. (2015), 'Pernicious Manipulation of Performance Measures in China's Cadre Evaluation System', *The China Quarterly*, 223, 618–37.

Garnaut, R. (2014), 'China's Role in Global Climate Change Mitigation', *China & World Economy*, 22 (5), 2–18.

Geels, F. W. (2014), 'Regime Resistance against Low-Carbon Transitions: Introducing Politics and Power into the Multi-Level Perspective', *Theory, Culture & Society*, 31 (5), 21–40.

Geels, F. W. and J. Schot (2007), 'Typology Of Sociotechnical Transition Pathways', *Research Policy*, 36 (3), 399–417.

Giddens, A. (1968), '"Power" in the Recent Writings of Talcott Parsons', *Sociology*, 2 (3), 257–72.

Giddens, A. (2009), *The Politics of Climate Change*, Cambridge: Polity Press.

Gilley, B. (2012), 'Authoritarian Environmentalism and China's Response to Climate Change', *Environmental Politics*, 21 (2), 287–307.

Goepe (2018), 'Carbon Trading Market: The Future Scale Is Expected to Exceed a Trillion Yuan', accessed 10 September 2019 at http://www.goepe.com/news/detail -375030.html.

Gordon, C. (1980), 'Afterword', in Colin Gordon (ed.), *Michel Foucault: Power/ Knowledge: Selected Interviews and Other Writings 1972–1977*, New York: Pantheon Books, pp. 229–59.

Gordon, C. (1991), 'Governmental Rationality: An Introduction', in G. Burchell, C. Gordon and P. Miller (eds.), *The Foucault Effect: Studies in Governmentality, with Two Lectures by and an Interview with Michel Foucault*, London: Harvest Wheatsheaf, pp. 1–51.

Green Finance Committee (GFC) (2018), 'The Third Meeting of the Chinese Working Group on the Climate and Environmental Information Disclosure Pilot of Chinese and British Financial Institutions Was Held in Huzhou on June 11', accessed 10 July 2018 at http://www.greenfinance.org.cn/displaynews.php?id=2182.

Green Finance Research Group (GFRG) of Industrial and Commercial Bank of China (ICBC) (2017), *ESG Green Rating and Green Index*, Beijing: ICBC.

Green Finance Task Force (GFTF) (2015), 'China Green Finance Task Force Report: Establishing China's Green Financial System', accessed 21 April 2020 at https:// unepinquiry.org/publication/establishing-chinas-green-financial-system/.

Green Low-Carbon Development Think-Tank Partnership (GLCDTTP) (ed.) (2016), *Policies and Practices in China's Low-Carbon Pilot Cities: Bring the Emissions Peak Forward*, Beijing: Science Press.

Green Sohu (2010), 'Ministry of Finance: The Scale of Special Fund for Energy Conservation, Emissions Reduction and New Energy Reaches 100 Billion Yuan', accessed 6 July 2020 at http://green.sohu.com/20100925/n275248896.shtml.

Green, F. and N. Stern (2015), 'China's "New Normal": Better Growth, Better Climate', accessed 17 May 2020 at http://www.lse.ac.uk/GranthamInstitute/wp-content/uploads/2015/05/Green-and-Stern-policy-paper-March-2015a.pdf.

Green, F. and N. Stern (2017), 'China's Changing Economy: Implications for its Carbon Dioxide Emissions', *Climate Policy*, 17 (4), 423–42.

Greenhalgh, S. (2005), 'Globalization and Population Governance in China', in A. Ong and S. J. Collier (eds.), *Global Assemblages: Technology, Politics, and Ethics as Anthropological Problems*, Malden, MA: Blackwell.

Greenhalgh, S. and E. A. Winckler (2005), *Governing China's Population: From Leninist to Neoliberal Biopolitics*, Stanford, CA: Stanford University Press.

Grin, J., J. Rotmans and J. Schot (2010), *Transitions to Sustainable Development: New Directions in the Study of Long Term Transformative Change*, London and New York: Routledge.

Gu, X.-L. (2016), 'Sharing Zhejiang's Experience of Carbon Measurement', accessed 9 May 2020 at http://diqiu.1she.com/8986/236354.html.

Hall, P. A. (1993), 'Policy Paradigms, Social Learning, and the State: The Case of Economic Policymaking in Britain', *Comparative Politics*, 25 (3), 275–96.

Hangzhou CCP Municipal Committee (HCCPMC) and Hangzhou Municipal People's Government (HMPG) (2009), 'Decision on Constructing a Low-Carbon City' (No. 37), accessed 11 May 2020 at http://www.doc88.com/p-796929165815.html.

Hangzhou Development and Reform Commission (HDRC) (2014), 'Hangzhou Climate Change Plan (2013–2020)', accessed 8 May 2020 at https://max.book118.com/html/2017/0711/121759734.shtm.

Hangzhou Municipal Bureau of Statistics (HMBS) (2016), 'Statistical Bulletin of Economic and Social Development 2015', accessed 20 May 2020 at http://www.hangzhou.gov.cn/art/2016/3/24/art_805865_663727.html.

Hangzhou Municipal Bureau of Statistics (HMBS) (2017), 'Statistical Bulletin of Economic and Social Development 2016', accessed 8 May 2020 at http://www.hangzhou.gov.cn/art/2017/3/10/art_812262_5885634.html.

Hangzhou Municipal Bureau of Statistics (HMBS) (2019), 'Statistical Bulletin of Economic and Social Development 20118', accessed 8 May 2020 at http://www.hangzhou.gov.cn/art/2019/3/4/art_805865_30593279.html.

Hangzhou Municipal People's Government (HMPG) (2011), 'Hangzhou's 12th Five Year Plan for Low Carbon Urban Development', accessed 20 April 2021 at http://dtfz.ccchina.org.cn/Detail.aspx?newsId=45742&TId=171.

Hangzhou Municipal People's Government (HMPG) (2016), 'The 13th Five-Year Planning Outline for the Socio-Economic Development of Hangzhou Municipality', accessed 29 June 2017 at http://www.hangzhou.gov.cn/art/2016/4/29/art_933506_690339.html.

Hangzhou Municipal People's Government (HMPG) (2017), 'Circular on the Implementation Plan of Controlling Greenhouse Gas Emissions in Hangzhou during the 13th Five Year Plan' (No. 172), accessed 12 November 2019 at http://www.hangzhou.gov.cn/art/2018/1/10/art_1450801_4228.html.

Hangzhou Municipal People's Government Office (HMPGO) (2013), 'Circular on the Administrative Measures of Allocating the Comprehensive Incentive Fund for Hangzhou as a Comprehensive Fiscal Policy Demonstration City for Energy Conservation and Emissions Reduction', accessed 2 December 2019 at http://law.esnai.com/mview/130162.

He, J.-K. (2020), 'A Study on the Long-Term Low-Carbon Development Strategy of China: Project Debriefing', accessed 7 April 2021 at https://www.efchina.org/News-zh/Program-Updates-zh/programupdate-lceg-20201015-zh.

Heilmann, S. (2008), 'Policy Experimentation in China's Economic Rise', *Studies in Comparative International Development*, 43 (1), 1–26.

Henderson, M. (2018), 'China's Environment: Views from Above, Below, and Beyond', *Global Environmental Politics*, 18 (1), 140–5.

Heyanyueche (2018), 'The Past and Present of License Plate Auction in Shanghai (Part 1)' [Serial], accessed 28 August 2020 at https://news.yiche.com/hao/wenzhang/788491.

Hood, C. (2011), *Summing up the Parts: Combining Policy Instruments for Least-Cost Climate Mitigation Strategies*, Paris: International Energy Agency. Downloadable from: https://www.iea.org/reports/summing-up-the-parts.

Hoogma, R. J. F., R. P. M. Kemp, J. Schot and B. Truffer (2002), *Experimenting for Sustainable Transport: The Approach of Strategic Niche Management*, London: Spon Press.

Hu, J. (2007), 'Hold High the Great Banner of Socialism with Chinese Characteristics and Strive for New Victories in Building a Moderately Prosperous Society in All Respects'. Report to the Seventeenth National Congress of the Communist Party of China on 15 October 2007, accessed 21 January 2018 at https://www.chinadaily.com.cn/china/2007-10/24/content_6204564.htm.

Hu, J. (2012), 'Firmly March on the Path of Socialism with Chinese Characteristics and Strive to Complete the Building of a Moderately Prosperous Society in All Respects'. Report to the Eighteenth National Congress of the Communist Party of China on 8 November 2012, accessed 21 January 2018 at http://language.chinadaily.com.cn/news/2012-11/19/content_15941774_3.htm.

Hu, M.-C., C.-Y. Wu and T. Shih (2015), 'Creating a New Socio-Technical Regime in China: Evidence from the Sino-Singapore Tianjin Eco-City', *Futures*, 70, 1–12.

Huang, P. and V. Castán Broto (2018), 'Interdependence between Urban Processes and Energy Transitions: The Dimensions of Urban Energy Transitions (DUET) Framework', *Environmental Innovation and Societal Transitions*, 28, 35–45.

Huang, P., V. Castán Broto, Y. Liu and H. Ma (2018), 'The Governance of Urban Energy Transitions: A Comparative Study of Solar Water Heating Systems in Two Chinese Cities', *Journal of Cleaner Production*, 180, 222–31.

Huang, P., S. O. Negro, M. P. Hekkert and K. Bi (2016), 'How China Became a Leader in Solar PV: An Innovation System Analysis', *Renewable and Sustainable Energy Reviews*, 64, 777–89.

Huang, Y. and X. Wang (2018), 'Strong on Quantity, Weak on Quality', in R. Garnaut, L. Song and C. Fang (eds.), *China's 40 Years of Reform and Development 1978–2018*, Canberra: ANU Press, pp. 291–312.

iCIBA (2008), 'Expert Translation: How to Correctly Translate "Shengtai Wenming"', accessed 8 March 2021 at http://news.iciba.com/study/oral/1192.shtml.

Industrial Bank Green Finance Group (IBGFG) (2018), *Commercial Banks' Exploration of Green Finance and Practices*, Beijing: China Finance Press.

Industrial and Commercial Bank of China (ICBC) (2017), *Impact of Environmental Factors on Credit Risk of Commercial Banks: Research and Application by ICBC Based on Stress Test*, Beijing: ICBC and Green Finance Committee.

Intergovernmental Panel on Climate Change (IPCC) (2007), *Summary for Policymakers of the Synthesis Report of the IPCC Fourth Assessment Report*, Cambridge: Cambridge University Press.

Intergovernmental Panel on Climate Change (IPCC) (2014), *Climate Change 2014: Synthesis Report*. Contribution of Working Groups I, II and III to the Fifth Assessment Report of the Intergovernmental Panel on Climate Change [Core Writing Team, R. K. Pachauri and L. A. Meyer (eds.)], Geneva: IPCC.

Intergovernmental Panel on Climate Change (IPCC) (2018), 'Global Warming of 1.5°C', Geneva. Downloadable from: https://www.ipcc.ch/sr15/.

International Energy Agency (IEA) (2007), *World Energy Outlook 2007: China and India Insights*, Paris: OECD.

International Energy Agency (IEA) (2018a), *World Energy Outlook 2017*, Paris: OECD.

International Energy Agency (IEA) (2018b), *Energy Efficiency 2018: Analysis and Outlooks to 2040*, accessed 28 March 2021 at https://www.iea.org/reports/energy-efficiency-2018.

International Energy Agency (IEA) (2019), *Global Energy & CO$_2$ Status Report: The Latest Trends in Energy and Emissions in 2018*, Paris: OECD.

International Energy Agency (IEA) (2021a), 'Press Release: Energy and Climate Leaders from US, China, EU, India and Other Key Economies to Boost Clean Energy Momentum at IEA-COP26 Net Zero Summit', accessed 15 March 2021 at https://www.iea.org/news/energy-and-climate-leaders-from-us-china-eu-india-and-other-key-economies-to-boost-clean-energy-momentum-at-iea-cop26-net-zero-summit.

International Energy Agency (IEA) (2021b), 'China. CO$_2$ Emissions Drivers, People's Republic of China 1990–2018, accessed 21 February 2021 at https://www.iea.org/countries/china.

International Institute of Green Finance (IIGF) and Research Centre for Climate and Energy Finance (RCCEF) (2017), *2017 China Climate Financing Report Briefing*, Beijing: IIGF and RCCEF, CUFE.

International Renewable Energy Agency (IRENA) (2019), *Renewable Power Generation Costs in 2018*, Abu Dhabi.

Jacobsson, S. and A. Johnson (2000), 'The Diffusion of Renewable Energy Technology: An Analytical Framework and Key Issues for Research', *Energy Policy*, 28 (9), 625–40.

Jahiel, A. R. (1997), 'The Contradictory Impact of Reform on Environmental Protection in China', *The China Quarterly*, 149, 81–103.

Jessop, B. (1998), 'The Rise of Governance and the Risks of Failure: The Case of Economic Development', *International Social Science Journal*, 50 (155), 29–45.

Jiang, Z. (1992), 'Accelerating the Reform, the Opening to the Outside World and the Drive for Modernization, so as to Achieve Greater Successes in Building Socialism With Chinese Characteristics', accessed 15 June 2020 at http://www.bjreview.com.cn/document/txt/2011-03/29/content_363504.htm.

Jiang, Z. (1997), 'Jiang Zemin's Report to the 15th Congress of the CPC on 12 Sept. 1997', accessed 21 January 2018 at http://cpc.people.com.cn/GB/64162/64168/64568/65445/4526290.html.

Jiang, Z. (2001), 'Jiang Zemin's Speech at the 80th Anniversary of the Founding of the CCP', accessed 17 January 2021 at http://www.cctv.com/special/1060/6/1.html.

Jingan District Development and Reform Committee (JDDRC) (2016), 'Implementation Opinions on Guiding Enterprises to Do Well in Energy Conservation and Emissions Reduction' (No. 15), Shanghai: Jingan District People's Government.

Kim, J.-Y. and L.-Y. Zhang (2008), 'Formation of Foreign Direct Investment Clustering: A New Path to Local Economic Development? The Case of Qingdao', *Regional Studies*, 42 (2), 265–80.

Knight, J. B. (2014), 'China as a Developmental State', *The World Economy*, 37 (10), 1335–47.

Kostka, G. and W. Hobbs (2012), 'Local Energy Efficiency Policy Implementation in China: Bridging the Gap between National Priorities and Local Interests', *The China Quarterly*, 211, 765–85.

Kung, J. K.-S., C. Xu and F. Zhou (2013), 'From Industrialization to Urbanization: The Social Consequences of Changing Fiscal Incentives on Local Governments' Behavior', in D. Kennedy and J. E. Stiglitz (eds.), *Law and Economics with Chinese Characteristics*, Oxford: Oxford University Press.

Lardy, N. R. (2008), 'Financial Repression in China', accessed 13 April 2020 at https://www.piie.com/publications/pb/pb08-8.pdf.

Lardy, N. R. (2019), 'State Sector Support in China Is Accelerating', accessed 13 April 2020 at https://www.piie.com/blogs/china-economic-watch/state-sector-support -china-accelerating.

Lemke, T. (2002), 'Foucault, Governmentality, and Critique', *Rethinking Marxism*, 14 (3), 49–64.

Li, V. and G. Lang (2010), 'China's "Green GDP" Experiment and the Struggle for Ecological Modernisation', *Journal of Contemporary Asia*, 40 (1), 44–62.

Li, W.-M., W.-J. Dong and Y. Qi (2017), 'A Review of Low Carbon Development during the 12th Five Year Plan Period', in X. Zhang and Y. Ji (eds.), *Annual Review of Low-Carbon Development in China*, Beijing: Social Sciences Academic Press (China), pp. 1–37.

Lieberthal, K. G. and D. M. Lampton (1992), *Bureaucracy, Politics, and Decision Making in Post-Mao China*, Berkeley: University of California Press.

Liu, H. W., D. Ding and H. Q. Xu (2015), 'Research Report on Hangzhou National Low Carbon City Pilot Project', accessed 7 June 2020 at http://www.ncsc.org.cn/yjcg/dybg/201501/t20150115_609605.shtml.

Liu, L., C. Chen, Y. Zhao and E. Zhao (2015), 'China's Carbon-Emissions Trading: Overview, Challenges and Future', *Renewable and Sustainable Energy Reviews*, 49, 254–66.

Liu, R. (2016), 'The Road to the Development of Human Ecological Civilization', in *Ecological Democracy*, trans. R. Liu, J. Zhang, and Y. Li, Beijing: China Environment Press.

Liu, S. and C. Zhang (2015), 'Foreword – Development Research Centre', in International Institute for Sustainable Development (IISD) and Development Research Center (DRC) of the State Council (eds.), *Greening China's Financial System*, Winnipeg: IISD, pp. vii–viii.

Lo, K. (2014), 'China's Low-Carbon City Initiatives: The Implementation Gap and the Limits of the Target Responsibility System', *Habitat International*, 42, 236–44.

Loorbach, D. (2010), 'Transition Management for Sustainable Development: A Prescriptive, Complexity-Based Governance Framework', *Governance*, 23 (1), 161–83.

Loorbach, D., N. Frantzeskaki and L. R. Huffenreuter (2015), 'Transition Management: Taking Stock from Governance Experimentation', *Journal of Corporate Citizenship*, 58, 48–66.

Lu, Z., L. Qian and Q. Fang (2019), 'Great Breakthrough in the Construction of Green Standards: A Commentary on the Green Industry Guidance Catalogue', accessed 27 April 2020 at http://greenfinance.xinhua08.com/a/20190307/1802344.shtml.

Ma, G., I. Roberts and G. Kelly (2018), 'China's Economic Rebalancing: Drivers, Outlook and the Role of Reform', in R. Garnaut, L. Song and C. Fang (eds.), *China's 40 Years of Reform and Development 1978–2018*, Canberra: ANU Press, pp. 187–213.

Ma, J. (2018), 'Ma Jun Explains Green Financial Re-Financing and Green Macro-Prudential Assessment (MPA)', accessed 26 April 2020 at http://greenfinance.xinhua08.com/a/20180401/1754795.shtml.

Ma, J. (2020), 'Work Report of the Green Finance Committee in 2019/20 and Outlook in 2020/21', accessed 9 March 2021 at http://www.greenfinance.org.cn/displaynews.php?id=2982.

Maffesoli, M. (1991), 'The Ethic of Aesthetics', *Theory, Culture & Society*, 8 (1), 7–20.

Mai, Q. and M. Francesch-Huidobro (2014), *Climate Change Governance in Chinese Cities*, London: Routledge.

Marinelli, M. (2018), 'How to Build a "Beautiful China" in the Anthropocene. The Political Discourse and the Intellectual Debate on Ecological Civilization', *Journal of Chinese Political Science*, 23 (3), 365–86.

Masiero, G., M. H. Ogasavara, A. C. Jussani and M. L. Risso (2016), 'Electric Vehicles in China: BYD Strategies and Government Subsidies', *RAI Revista de Administração e Inovação*, 13 (1), 3–11.

Mazzucato, M. (2018), *The Entrepreneurial State: Debunking Public vs. Private Sector Myths*, London: Penguin.

Mazzucato, M. and G. Semieniuk (2018), 'Bridging the Gap: The Role of Innovation Policy and Market Creation', in *Emission Gap Report 2018*, Nairobi, pp. 52–9.

McDowall, W., P. Ekins, S. Radošević and L. Zhang (2013), 'The Development of Wind Power in China, Europe and the USA: How Have Policies and Innovation System Activities Co-Evolved?', *Technology Analysis & Strategic Management*, 25 (2), 163–85.

Meadowcroft, J. (2005), 'From Welfare State to Ecostate', in J. Barry and R. Eckersley (eds.), *The State and the Global Ecological Crisis*, Cambridge, MA: MIT Press, pp. 3–32.

Meadowcroft, J. (2009), 'What about the Politics? Sustainable Development, Transition Management, and Long Term Energy Transitions', *Policy Sciences*, 42 (4), 323–40.

Meadowcroft, J. (2011), 'Engaging with the Politics of Sustainability Transitions', *Environmental Innovation and Societal Transitions*, 1 (1), 70–5.

Mertha, A. (2009), '"Fragmented Authoritarianism 2.0": Political Pluralization in the Chinese Policy Process', *The China Quarterly*, 200, 995–1012.

Miller, P. and N. Rose (1990), 'Governing Economic Life', *Economy and Society*, 19 (1), 1–31.

Ministry of Ecology and Environment (MEE) (2018a), 'China's Policy and Action in Addressing Climate Change (2018)'. Downloadable from: http://www.mee.gov.cn/ywgz/ydqhbh/qhbhlf/201811/P020181129539211385741.pdf.

Ministry of Ecology and Environment (MEE) (2018b), 'The People's Republic of China Second Biennial Update Report on Climate Change', Beijing. Downloadable from: https://unfccc.int/gcse?q=China%27s%20second%20bienniel%20report.

Ministry of Environment and Ecology (MEE) (2020), 'Notice on Disseminating the Implementation Plan for Setting and Allocating the Total Amount of National Carbon Emissions Trading Quota in 2019–2020 (Power Generation Industry) and the List of Key Emission Units Included in the National Carbon Emissions Trade', accessed 3 April 2021 at https://www.mee.gov.cn/xxgk2018/xxgk/xxgk03/202012/t20201230_815546.html.

Ministry of Environmental Protection (MEP) (2010), 'Circular on Public Consultation of Environmental Information Disclosure Guidelines of Listed Companies', accessed 24 August 2020 at http://www.gov.cn/gzdt/2010-09/14/content_1702292.htm.

Ministry of Finance (MoF) and Ministry of Transport (MoT) (2011), 'Interim Measures for the Administration of Special Funds for Transportation Energy Conservation and Emissions Reduction', accessed 25 March 2021 at http://www.gov.cn/zwgk/2011-07/06/content_1900241.htm.

Ministry of Industry and Information Technology (MIIT) (2016), 'Administrative Measures for Industrial Energy Conservation Management, China', accessed 4 January 2021 at http://www.gov.cn/gongbao/content/2016/content_5097562.htm.

Ministry of Industry and Information Technology (MIIT), National Development and Reform Commission (NDRC) and National Energy Administration (NEA) (2012), 'Action Program to Address Climate Change in Industry (2012–2020)', accessed 10 April 2020 at http://www.miit.gov.cn/n1146290/n1146417/n1146532/c3303711/content.html.

Ministry of Transport (MoT) (2012), 'Circular on Undertaking Low-Carbon Transport-System Development Second-Batch Pilots', accessed 13 May 2020 at http://guoqing.china.com.cn/zwxx/2012-02/08/content_24582654.htm.

Morrison, R. (1995), *Ecological Democracy*, Boston, MA: South End Press.

Nahm, J. (2017), 'Exploiting the Implementation Gap: Policy Divergence and Industrial Upgrading in China's Wind and Solar Sectors', *The China Quarterly*, 231, 705–27.

National Bureau of Statistics (NBS) (2015), 'Regulations on Statistical and Reporting Operations of Departments Involved in Addressing Climate Change (Trial)', accessed 20 March 2020 at http://www.stats.gov.cn/tjsj/tjzd/gjtjzd/201701/t20170109_1451402.html.

National Bureau of Statistics (NBS) (2019), 'China Statistics Yearbook 2018', accessed 13 April 2020 at http://www.stats.gov.cn/tjsj/ndsj/2018/indexeh.htm.

National Bureau of Statistics (NBS) (2021), 'Statistical Bulletin of the People's Republic of China on National Economic and Social Development in 2020', accessed 6 April 2021 at http://www.stats.gov.cn/tjsj/zxfb/202102/t20210227_1814154.html.

National Development and Reform Commission (NDRC) (2004), 'Medium- and Long-Term Special Plan for Energy Conservation', accessed 28 February 2021 at https://www.ndrc.gov.cn/fggz/hjyzy/jnhnx/200507/t20050711_1135117.html.

National Development and Reform Commission (NDRC) (2005), 'NDRC Launching Ten Key Energy Conservation Projects', accessed 28 February 2021 at https://www.ndrc.gov.cn/fggz/hjyzy/jnhnx/200507/t20050711_1135126.html.

National Development and Reform Commission (NDRC) (2007a), 'China's National Climate Change Programme', accessed 1 April 2020 at https://sustainabledevelopment.un.org/index.php?page=view&type=99&nr=61&menu=1449.

National Development and Reform Commission (NDRC) (2007b), 'Implementation Scheme of Energy Consumption per Unit GDP Assessment System', accessed 11 April 2020 at http://www.gov.cn/zwgk/2007-11/23/content_813617.htm.

National Development and Reform Commission (NDRC) (2007c), 'The Medium- and Long-Term Development Planning of Renewable Energy', accessed 12 August 2019 at http://www.ndrc.gov.cn/fzgggz/hjbh/jnjs/200507/t20050711_45823.html.

National Development and Reform Commission (NDRC) (2010), 'Circular of the National Development and Reform Commission on the Pilot Work of Low Carbon

Provinces and Cities' (No. 1587), accessed 07 June 2020 at http://www.gov.cn/zwgk/2010-08/10/content_1675733.htm.

National Development and Reform Commission (NDRC) (2011a), 'Guidelines for the Compilation of Provincial Greenhouse Gas Inventories (Trial)', accessed at http://www.cbcsd.org.cn/sjk/nengyuan/standard/home/20140113/download/shengjiwenshiqiti.pdf.

National Development and Reform Commission (NDRC) (2011b), 'Circular on Disseminating the Implementation Plan of Energy Conservation and Low-Carbon Action of Ten Thousand Enterprises', accessed 28 February 2021 at https://www.ndrc.gov.cn/xxgk/zcfb/tz/201112/t20111229_964360.html.

National Development and Reform Commission (NDRC) (2011c), 'Development Planning for Shandong Peninsula Blue Economy Economic Region', accessed 8 May 2020 at https://max.book118.com/html/2019/0421/5014030022002031.shtm.

National Development and Reform Commission (NDRC) (2012), 'Second National Communication on Climate Change of The People's Republic of China', NDRC, Beijing. Downloadable from: https://unfccc.int/resource/docs/natc/chnnc2e.pdf.

National Development and Reform Commission (NDRC) (2014a), 'National Plan on Climate Change (2014–2020)' (No. 2347), accessed 4 January 2019 at https://www.ndrc.gov.cn/xxgk/zcfb/tz/201411/t20141104_963642.html.

National Development and Reform Commission (NDRC) (2014b), 'A Letter on the Trial Implementation of Strengthening the Risk Prevention of Enterprise Bonds in an All-Round Way', accessed 26 April 2020 at http://blog.sina.com.cn/s/blog_13ba93ba60102v4ca.html.

National Development and Reform Commission (NDRC) (2015), 'Green Bond Issuance Guidelines' (No. 3504), accessed 7 July 2018 at http://www.ndrc.gov.cn/zcfb/zcfbtz/201601/t20160108_770871.html.

National Development and Reform Commission (NDRC) (2016), 'The People's Republic of China First Biennial Update Report on Climate Change'. Downloadable from: https://unfccc.int/sites/default/files/resource/3_China_FSV_Presentation.pdf.

National Development and Reform Commission (NDRC) (2017a), 'The Current State of the Compilation of Provincial Greenhouse Gas Inventories', accessed 27 August 2020 at http://www.tanpaifang.com/tanpancha/2017/0807/60227.html.

National Development and Reform Commission (NDRC) (2017b), 'Circular on the Pilot Work of Third-Batch Low Carbon Cities' (No. 66), accessed 29 November 2019 at http://www.ndrc.gov.cn/zcfb/zcfbtz/201701/t20170124_836394.html.

National Development and Reform Commission (NDRC) (2021), 'Report on the Implementation of National Economic and Social Development Plan 2020 and the Draft National Economic and Social Development Plan 2021', accessed 24 March 2021 at http://www.gov.cn/xinwen/2021-03/13/content_5592786.htm.

National Development and Reform Commission (NDRC), Ministry of Industry and Information Technology (MIIT), Ministry of Natural Resources (MNR), Ministry of Environment and Ecology (MEE), Ministry of Housing and Urban and Rural Development (MHURD), The People's Bank of China (PBC) and National Energy Administration (NEA) (2019), 'Circular on Disseminating the Green Industry Guidance Catalogue (2019 Edition)' (No. 293), accessed 17 February 2021 at https://www.ndrc.gov.cn/fggz/hjyzy/stwmjs/201903/t20190305_1220625.html.

National Development and Reform Commission (NDRC), National Bureau of Statistics (NBS), Ministry of Environmental Protection (MEP) and CCPCC Organization Department (CCPCCOD) (2016), 'Circular on Green Development Indicators System and the Assessment Target System for Constructing Ecological Civilization', accessed 11 September 2020 at http://www.gov.cn/xinwen/2016-12/22/content _5151575.htm.

National Development and Reform Commission (NDRC) and National Energy Administration (NEA) (2016a), 'Energy Production and Consumption Revolution Strategy', accessed 2 April 2020 at https://www.ndrc.gov.cn/xxgk/zcfb/tz/201704/ t20170425_962953.html.

National Development and Reform Commission (NDRC) and National Energy Administration (NEA) (2016b), 'The 13th Five-Year Planning for Energy Development', accessed 5 January 2021 at https://www.ndrc.gov.cn/xxgk/zcfb/tz/ 201701/t20170117_962873.html.

National Development and Reform Commission (NDRC) and National Energy Administration (NEA) (2018), 'Clean Energy Consumption Action Plan (2018–2020)' (No. 1575), accessed 28 March 2021 at https://www.ndrc.gov.cn/ xxgk/zcfb/ghxwj/201812/t20181204_960958.html%0A.

National Development and Reform Commission (NDRC) and National Energy Administration (NEA) (2019), 'Circular on the Establishment and Improvement of Renewable Energy Power Consumption Guarantee Mechanism', accessed 28 March 2021 at http://zfxxgk.nea.gov.cn/auto87/201905/t20190515_3662.htm.

National Development and Reform Commission (NDRC) and National Energy Administration (NEA) (2020), 'Circular on Disseminating the Weightings of Renewable Energy Electricity Consumption Responsibility in 2020 in Provincial Administrative Regions', accessed 1 April 2021 at https://www.ndrc.gov.cn/xxgk/ zcfb/tz/202006/t20200601_1229674.html.

National Energy Administration (NEA) (2016), 'Guiding Opinions on Establishing a Target Guiding System for Renewable Energy Development and Utilization', accessed 27 March 2021 at http://zfxxgk.nea.gov.cn/auto87/201603/t20160303 _2205.htm.

National Energy Administration (NEA) (2020), 'Monitoring and Evaluative Bulletin of Renewable Energy Development in China in 2019', accessed 28 March 2021 at http://www.gov.cn/zhengce/zhengceku/2020-05/16/content_5512148.htm.

National People's Congress (NPC) (1954), 'Constitution of the People's Republic of China (1954)', accessed 22 March 2020 at http://en.pkulaw.cn/display.aspx?cgid= 52993&lib=law.

Network for Greening the Financial System (NGFS) (2017), 'First Meeting of the Central Banks and Supervisors Network for Greening the Financial System (NGFS) on January 24th in Paris', accessed 22 April 2020 at https://www.banque-france .fr/sites/default/files/medias/documents/press-release_2018-01-26_first-meeting-of -the-central-banks-and-supervisors-network-for-greening-the-financial-system-ngfs .pdf.

Network for Greening the Financial System (NGFS) (2019), 'First Comprehensive Report: A Call for Action', accessed 29 April 2020 at https://www.ngfs.net/en/liste -chronologique/ngfs-publications?year=2019.

Newell, P., H. Bulkeley, K. Turner, C. Shaw, S. Caney, E. Shove and N. Pidgeon (2015), 'Governance Traps in Climate Change Politics: Re-framing the Debate in Terms of Responsibilities and Rights', *Wiley Interdisciplinary Reviews: Climate Change*, 6 (6), 535–40.

Nordhaus, W. D. (2006a), 'After Kyoto: Alternative Mechanisms to Control Global Warming', *American Economic Review*, 96 (2), 31–4.

Nordhaus, W. D. (2006b), *The 'Stern Review' on the Economics of Climate Change*, Working Paper 12741, Cambridge, MA: National Bureau of Economic Research. Downloadable from: http://www.nber.org/papers/w12741.

Oels, A. (2005), 'Rendering Climate Change Governable: From Biopower to Advanced Liberal Government?', *Journal of Environmental Policy & Planning*, 7 (3), 185–207.

Okereke, C., H. Bulkeley and H. Schroeder (2009), 'Conceptualizing Climate Governance Beyond the International Regime', *Global Environmental Politics*, 9 (1), 58–78.

Pan, J. (2018), 'The Evolution and Transformation of China's Climate Change Response Strategy: From Preventing "Black Swan" Events to Reducing "Grey Rhino" Risk', in R. Garnaut, L. Song, and C. Fang (eds.), *China's 40 Years of Reform and Development 1978–2018*, Canberra: ANU Press, pp. 525–42.

Pan, Y. (2003), 'Environmental Culture and National Rejuvenation', Keynote Speech at the First 'Green China Forum', accessed 20 August 2020 at https://web.archive.org/web/20070821152956/http://env.people.com.cn/GB/8220/50110/3502857.html.

Pan, Y. (2004), 'Environmental Protection Indicators and Achievement Assessment for Officials', *World Environment*, 3, 6–9.

Parsons, T. (1963), 'On the Concept of Political Power', *Proceedings of the American Philosophical Society*, 107 (3), 232–62.

Paterson, M. and J. Stripple (2010), 'My Space: Governing Individuals' Carbon Emissions', *Environment and Planning D: Society and Space*, 28 (2), 341–62.

Peng, H. W. (2020), '1975: How Did Deng Xiaoping Skilfully Negate "Taking Class Struggle as the Key Link"?', accessed 9 March 2021 at http://wenhui.whb.cn/zhuzhan/dushu/20200508/346039.html.

People's Bank of China (PBC) (1995), 'Circular on Implementing Credit Policy and Strengthening the Work on Environmental Protection' (No. 24), accessed 28 February 2021 at https://www.chinaacc.com/new/63/69/110/1995/2/ad594249401116259919555.htm.

People's Bank of China (PBC) (2015), 'Announcement on Matters Concerning the Issue of Green Financial Bonds in the Inter-Bank Bond Market' (No. 39), accessed 26 April 2020 at http://www.gov.cn/xinwen/2015-12/22/content_5026636.htm.

People's Bank of China (PBC) (2017), *2017 Annual Report*, Beijing.

People's Bank of China (PBC) (2018), 'Circular on the Establishment of a Green Credit Special Statistical System' (No. 180), accessed 26 April 2020 at http://greenfinance.xinhua08.com/a/20180727/1770939.shtml.

People's Bank of China (PBC) (2019), 'Statistical Report on Credit Structure of Financial Institutions, 2018', accessed 27 April 2020 at http://www.pbc.gov.cn/en/3688247/3688978/3709143/3757386/index.html.

People's Bank of China (PBC) (2021), 'G20 Sustainable Finance Study Group Resumed', accessed 8 March 2021 at http://www.greenfinance.org.cn/displaynews.php?cid=21&id=3156.

People's Bank of China (PBC) and China Securities Regulatory Commission (CSRC) (2017), 'Interim Guidelines for Green Bond Assessment and Certification' (No. 20), accessed 24 August 2020 at http://www.pbc.gov.cn/rmyh/105208/3449893/index.html.

People's Bank of China (PBC), Ministry of Finance (MoF), National Development and Reform Commission (NDRC), Ministry of Environmental Protection (MEP), China Banking Regulatory Commission (CBRC), China Securities Regulatory Commission (CSRC) and China Insurance Regulatory Commission (CIRC) (2016), 'Guidelines for Establishing the Green Financial System' (No. 228), Beijing, accessed 14 March 2017 at http://www.scio.gov.cn/32344/32345/35889/36819/xgzc36825/Document/1555348/1555348.htm.

People's Bank of China (PBC), National Development and Reform Commission (NDRC) and China Securities Regulatory Commission (CSRC) (2020), 'Green Bond Endorsed Projects Catalogue (2020 Version)' (Consultation Draft). Downloadable from: http://www.pbc.gov.cn/tiaofasi/144941/144979/3941920/4052500/index .html.

People's Bank of China Research Bureau (PBCRB) (2018), *China Green Finance Progress Report 2017*, Beijing: China Financial Publishing House.

Peters, B. G. and J. Pierre (1998), 'Governance without Government? Rethinking Public Administration', *Journal of Public Administration Research and Theory*, 8 (2), 223–43.

Ploberger, C. (2013), 'China's Adaptation Challenges: A Critical Assessment of China's Ability to Facilitate a Strategic Shift towards a Low-Carbon Economy by Applying the Structure–Agency Framework', *Journal of Contemporary China*, 22 (84), 1028–47.

Policy Research Center on Environment and Economy (PRCEE) (2019), 'Research Report on Public Low Carbon Lifestyle in the Context of Internet Platforms', Beijing. Downloadable from: http://www.prcee.org/yjcg/yjbg/201909/t20190909 _733041.html.

Popp, D. (2010), 'Innovation and Climate Policy', *Annual Review of Resource Economics*, 2 (1), 275–98.

PricewaterhouseCoopers (PwC) (2013), *Exploring Green Finance Incentives in China*. Final Report. Shenzhen, PWC, accessed 13 April 2020 at https://www.pwchk.com/en/migration/pdf/green-finance-incentives-oct2013-eng.pdf.

PricewaterhouseCoopers (PwC) (2017), *The Low Carbon Economy Index 2017: Is Paris Possible?*, accessed 17 May 2020 at https://www.pwc.co.uk/sustainability -climate-change/assets/pdf/lcei-17-pdf-final-v2.pdf.

PricewaterhouseCoopers (PwC) (2018), *The Low Carbon Index 2018: Time to Get on with It*, accessed 6 December 2019 at https://www.pwc.co.uk/sustainability-climate -change/assets/pdf/low-carbon-economy-index-2018-final.pdf.

PricewaterhouseCoopers (PwC) (2019), *The Low Carbon Economy Index 2019*, accessed 1 April 2020 at https://www.pwc.co.uk/services/sustainability-climate -change/insights/low-carbon-economy-index.html.

PricewaterhouseCoopers (PwC) (2020), *Net Zero Economy Index 2020: The Pivotal Decade*, London. Downloadable from: https://www.pwc.co.uk/services/sustainability-climate-change/insights/net-zero-economy-index.html.

Pronina, L. (2019), 'What Are Green Bonds and How "Green" is Green?', accessed 21 April 2020 at https://www.bloomberg.com/news/articles/2019-03-24/what-are -green-bonds-and-how-green-is-green-quicktake.

Qi, Y. (2018), 'Introduction', in Y. Qi and X. Zhang (eds.), *Annual Review of Low-Carbon Development in China*, Beijing: Social Sciences Academic Press (China), pp. 1–6.

Qi, Y., L. Ma, H. Zhang and H. Li (2008), 'Translating a Global Issue Into Local Priority: China's Local Government Response to Climate Change', *The Journal of Environment & Development*, 17 (4), 379–400.

Qingdao Development and Reform Commission (QDRC) (2014), 'Circular on Organizing the Implementation of Qingdao Low Carbon Development Plan (2014-2020)', Qingdao, accessed 11 May 2020 at http://www.qingdao.gov.cn/n172/n68422/n68424/n30259215/n30259219/140924163931863706.html.

Qingdao Development and Reform Commission (QDRC) (2017), 'Announcement on 2017 Central Budget Investment Plan (the First Batch) for Key Projects of Resource Conservation and Recycling', accessed 9 March 2021 at http://dpc.qingdao.gov.cn/n32569055/200102144553130665.html.

Qingdao Development and Reform Commission (QDRC) (2018), 'Our City Successfully Completed the Control Target of Greenhouse Gas Emissions in 2017', accessed 14 June 2020 at http://dpc.qingdao.gov.cn/n32569057/n32569083/200108104130442717.html.

Qingdao Development and Reform Commission (QDRC) (2019), 'Deepening the Pilot Project of Low Carbon City and Promote the Transformation and Upgrading of Green City', accessed 5 November 2019 at http://dpc.qingdao.gov.cn/n32205328/n32205340/n32205344/190611183422408373.html.

Qingdao Municipal Bureau of Statistics (QMBS) (2019), '2018 Statistical Bulletin of Economic and Social Development of Qingdao', accessed 8 May 2020 at http://qdtj.qingdao.gov.cn/n28356045/n32561056/n32561072/190319133354050380.html.

Qingdao Municipal People's Government (QMPG) (2011), 'The 12th Five-Year Plan for Qingdao's Economic and Social Development (Outline)' (No. 19), accessed 30 March 2020 at http://dpc.qingdao.gov.cn/.

Qingdao Municipal People's Government (QMPG) (2016a), 'The 13th Five-Year Plan for Qingdao's Economic and Social Development (Outline)' (No. 9), accessed 30 March 2020 at http://dpc.qingdao.gov.cn/.

Qingdao Municipal People's Government (QMPG) (2016b), 'The City of Qingdao Has Successfully Passed the Provincial Government Responsibility Target Assessment in 2015 and Exceeded the 12th Five Year Plan Energy Saving Target and Was in the Forefront of the Province', accessed 5 November 2019 at http://dpc.qingdao.gov.cn/n32205328/n32205340/n32205344/160531100828594245.html.

Qingdao Municipal People's Government (QMPG) (2018a), 'Circular on Qingdao Work Plan for Controlling the Emissions of Greenhouse Gases' (No. 8), accessed 9 May 2020 at http://www.qingdao.gov.cn/n172/n68422/n68424/n31282492/n31282495/180316111146623134.html.

Qingdao Municipal People's Government (QMPG) (2018b), 'Qingdao Action Programme for Energy Conservation and Green Development (2018–2020)' (No. 75), accessed 8 January 2020 at http://www.china-nengyuan.com/news/128267.html.

Qingdao Municipal Planning Bureau (QMPB) (2009), 'Qingdao City Master Plan (2006–2020) (Consultation Version)', accessed 15 March 2020 at http://www.cityup.org/case/general/20090424/47754-10.shtml.

Qingdao Rural Commercial Bank (QRCB) (2018), 'Quarterly Report on the Use of Green Bond Proceeds (2018 Q3)', accessed 9 March 2021 at http://file.finance.sina.com.cn/211.154.219.97:9494/MRGG/BOND/2018/2018-10/2018-10-26/9178978.PDF.

Qingdao West Coast Development Group (QWCDG) (2017), 'The Sino French Water Project of New Port City Company Received 17.8 Million Yuan from the Central Budget', accessed 10 March 2020 at http://www.xhafz.com/pc/content.html#/id=760.

Ramo, J. C. (2004), *The Beijing Consensus*, London: The Foreign Policy Centre.

Research Centre for Climate and Energy Finance (RCCEF) (2015), *2015 China Climate Financing Report*, Beijing: Central University of Finance and Economics.

Research Centre for Climate and Energy Finance (RCCEF) (2017), *2016 China Climate Financing Report*, Beijing: Central University of Finance and Economics.

Rice, J. L. (2010), 'Climate, Carbon, and Territory: Greenhouse Gas Mitigation in Seattle, Washington', *Annals of the Association of American Geographers*, 100 (4), 929–37.

Rose, N. (1999), *Powers of Freedom: Reframing Political Thought*, Cambridge: Cambridge University Press.

Rose, N. and P. Miller (1992), 'Political Power beyond the State : Problematics of Government', *The British Journal of Sociology*, 43 (2), 173–205.

Rotmans, J., R. Kemp and M. van Asselt (2001), 'More Evolution than Revolution: Transition Management in Public Policy', *Foresight*, 3 (1), 15–31.

Rutland, T. and A. Aylett (2008), 'The Work of Policy: Actor Networks, Governmentality, and Local Action on Climate Change in Portland, Oregon', *Environment and Planning D: Society and Space*, 26 (4), 627–46.

Savage, S. P. (1978), 'Political Power and Political Subsystems: Parsons' Analysis of Politics', *Economy and Society*, 7 (2), 157–74.

Schmidt, V. A. (2008), 'Discursive Institutionalism: The Explanatory Power of Ideas and Discourse', *Annual Review of Political Science*, 11 (1), 303–26.

Schoenhals, M. (1991), 'The 1978 Truth Criterion Controversy', *The China Quarterly*, 126, 243–68.

Senellart, M. (2008), 'Course Context', in M. Foucault, *The Birth of Biopolitics: Lectures at the Collège de France, 1978–1979*, Basingstoke: Palgrave Macmillan, pp. 327–30.

Shambaugh, D. (2013), *China Goes Global: The Partial Power*, Oxford: Oxford University Press.

Shanghai Development and Reform Commission (SDRC) (2019), '2018 Assessment Circular on the Target Responsibility System of Energy Conservation and Emissions Reduction of District Governments', accessed 1 May 2020 at http://www.shanghai .gov.cn/nw2/nw2314/nw2319/nw12344/u26aw62820.html.

Shanghai Development and Reform Commission (SDRC), Shanghai Economy and Information Technology Commission (SHEITC) and Shanghai Municipal Bureau of Finance (SMBF) (2014), 'Measures of Shanghai Municipality on the Implementation and Management of Differential Power Prices for Promoting the Adjustment of Industrial Structure', accessed 29 March 2020 at http://www.czj.sh.gov.cn/zys _8908/zcfg_8983/zcfb_8985/qy_9027/qt_9032/201509/t20150916_156810.shtml.

Shanghai Economy and Information Technology Commission (SHETIC) (2014), 'Guidance Catalogue for Production Technology, Equipment and Products of Restricted and Eliminated Industries in Shanghai (the First Batch)', accessed 29 March 2020 at http://www.shcpo.com.cn/index.php/law/shswbjwj/16-2014-09-17 -06-46-30.

Shanghai Economy and Information Technology Commission (SHETIC) (2020), 'Guiding Catalogue for the Adjustment of Industrial Structure: Restricted and Eliminated Types (2020 Edition) (Consultation Copy)', accessed 29 March 2020 at http://www.sheitc.sh.gov.cn/cyfz/684609.htm.

Shanghai Economy and Information Technology Commission (SHETIC), Shanghai Municipal Bureau of Statistics (SMBS), Shanghai Municipal Energy Conservation Supervisory Center and Shanghai Municipal Energy Efficiency Center (2014), 'The Negative List of the Adjustment of Industrial Structure and Guidelines for Energy Efficiency in Shanghai' (2014 Edition).

Shanghai Environment and Energy Exchange (SEEE) (2019), *2018 Shanghai Carbon Market Report*.

Shanghai Municipal Bureau of Statistics (SMBS) (2011), *Statistical Bulletin of Economic and Social Development 2010*, accessed 8 May 2020 at http://www.tjcn .org/tjgb/09sh/23580_6.html.

Shanghai Municipal Bureau of Statistics (SMBS) (2016), *Statistical Bulletin of Economic and Social Development 2015*, accessed 8 May 2020 at http://www.tjcn .org/tjgb/09sh/32640.html.

Shanghai Municipal Bureau of Statistics (SMBS) (2018), *Shanghai Statistical Yearbook 2017*, accessed 16 May 2019 at http://tjj.sh.gov.cn/html/sjfb/201803/1001690.html.

Shanghai Municipal Bureau of Statistics (SMBS) (2019), *Statistical Bulletin of Economic and Social Development 2018*, accessed 8 May 2020 at http://www.tjcn .org/tjgb/09sh/35767_3.html.

Shanghai Municipal Leading Group on Climate Change and Energy Conservation and Emissions Reduction (SMLGCCECER) (2019), 'Circular on 2019 Key Work Arrangement for Energy Conservation, Emissions Reductions and Climate Change', accessed 7 June 2020 at http://www.china-nengyuan.com/news/138008.html.

Shanghai Municipal People's Government (SMPG) (2016a), 'The 13th Five-Year Plan for Shanghai's Economic and Social Development', accessed 29 June 2017 at http:// www.shanghai.gov.cn/nw2/nw2314/nw2319/nw22396/nw39378/u21aw1101146 .html.

Shanghai Municipal People's Government (SMPG) (2016b), 'Shanghai City Master Plan (2016–2040)', accessed 25 November 2019 at https://baike.baidu.com/item/上 海市城市总体规划%282016-2040%29/19918887?fr=aladdin.

Shanghai Municipal People's Government (SMPG) (2017a), 'Circular on Issuing the 13th Five-Year Plan for Energy Conservation and Addressing Climate Change for Shanghai' (No. 12), accessed 27 November 2019 at http://www.shanghai.gov.cn/ nw2/nw2314/nw2319/nw22396/nw22403/u26aw51762.html.

Shanghai Municipal People's Government (SMPG) (2017b), 'Circular on Shanghai's 13th Five-Year Planning for Energy Development', accessed 12 June 2020 at http:// www.shanghai.gov.cn/nw2/nw2314/nw2319/nw12344/u26aw51932.html.

Shanghai Municipal People's Government (SMPG) (2018), 'Circular on Issuing the Comprehensive Work Plan for Energy Conservation and Controlling Greenhouse Gas Emissions in the 13th Five Year Plan of Shanghai', accessed 9 May 2020 at http://www.shanghai.gov.cn/nw2/nw2314/nw2319/nw12344/u26aw56058.html.

Shanghai Pudong New Area People's Government (SPNAPG) (2021), *Pudong Yearbooks*, accessed 9 March 2021 at http://www.pudong.gov.cn/shpd/ IntoYearbooks/?categorynum=008006033&type=1.

Shapiro, J. (2001), *Mao's War Against Nature: Politics and the Environment in Revolutionary China*, Cambridge: Cambridge University Press.

Shen, Y. and A. L. Ahlers (2018), 'Local Environmental Governance Innovation in China: Staging "Triangular Dialogues" for Industrial Air Pollution Control', *Journal of Chinese Governance*, 3 (3), 351–69.

Shenzhen Stock Exchange (SZSE) (2018), 'Answers to Questions Related to Green Corporate Bonds of Shenzhen Stock Exchange', accessed 11 February 2020 at https://wenku.baidu.com/view/bd45f1faa55177232f60ddccda38376bae1fe053 .html.

Sigley, G. (1996), 'Governing Chinese Bodies: The Significance of Studies in the Concept of Governmentality for the Analysis of Government in China', *Economy and Society*, 25 (4), 457–82.

Sigley, G. (2006), 'Chinese Governmentalities: Government, Governance and the Socialist Market Economy', *Economy and Society*, 35 (4), 487–508.

Smith, A., A. Stirling and F. Berkhout (2005), 'The Governance of Sustainable Socio-Technical Transitions', *Research Policy*, 34 (10), 1491–510.

Srivastav, S., S. Fankhauser and A. Kazaglis (2018), 'Low-Carbon Competitiveness in Asia', *Economies*, 6 (1). Downloadable from: https://doi.org/10.3390/economies6010005.

State Council (1981), 'Decision of the State Council on Strengthening Environmental Protection during the Period of National Economic Adjustment' (No. 4), accessed 24 April 2020 at http://www.law-lib.com/law/law_view.asp?id=44459.

State Council (1986), 'Interim Regulations on Energy Conservation Management', accessed 5 April 2020 at https://baike.baidu.com/item/节约能源管理暂行条例/7130774?fr=aladdin.

State Council (2005), 'Decision of the State Council on Implementing the Scientific Outlook on Development and Strengthening Environmental Protection' (No. 39), Beijing, accessed 5 March 2021 at http://www.gov.cn/zwgk/2005-12/13/content_125736.htm.

State Council (2006a), 'Decision on Strengthening Energy Conservation' (No. 28), accessed 5 April 2020 at http://www.gov.cn/zwgk/2006-08/23/content_368136.htm.

State Council (2006b), 'Outline of the 11th Five Year Plan for National Economic and Social Development of the People's Republic of China' (No. 12), accessed 5 April 2020 at http://www.gov.cn/gongbao/content/2006/content_268766.htm.

State Council (2007a), 'Regulation on the Disclosure of Government Information' (No. 492).

State Council (2007b), 'Circular of the Implementation Schemes and Methods of Statistical Monitoring and Assessment of Energy Conservation and Emissions Reduction by the State Council' (No. 36), accessed 12 April 2020 at http://www.gov.cn/zwgk/2007-11/23/content_813617.htm.

State Council (2007c), 'Comprehensive Work Plan for Energy Conservation and Emissions Reduction' (No. 15), accessed 12 April 2020 http://www.gov.cn/zwgk/2007-06/03/content_634545.htm.

State Council (2009), 'Circular on Adjusting the Capital Ratio of Fixed Asset Investment Projects' (No. 27), accessed 24 April 2020 at http://www.gov.cn/zwgk/2009-05/27/content_1326017.htm.

State Council (2010), 'Decision on Accelerating the Cultivation and Development of Strategic Emerging Industries' (No. 32), accessed 20 July 2020 at http://www.gov.cn/zhengce/content/2010-10/18/content_1274.htm.

State Council (2011a), 'Comprehensive Work Plan of Energy Conservation and Emissions Reduction in the 12th FYP Period' (No. 26), accessed 9 April 2020 at http://www.gov.cn/zwgk/2011-09/07/content_1941731.htm.

State Council (2011b), 'The 12th Five Year Plan for National Economic and Social Development of the People's Republic of China (Outline)', accessed 9 April 2020 at http://www.gov.cn/2011lh/content_1825838.htm.

State Council (2011c), 'The Work Plan for Controlling Greenhouse Gas Emissions in the 12th FYP Period' (No. 41), accessed 9 April 2020 at http://www.gov.cn/zwgk/2012-01/13/content_2043645.htm.

State Council (2012), 'Circular on Disseminating the National Strategic Emerging Industry Development Plan in the 12th Five Year Plan' (No. 28), accessed 19 July 2020 at http://www.gov.cn/zwgk/2012-07/20/content_2187770.htm.

State Council (2013), 'Circular on Issuing an Action Plan for the Prevention and Control of Air Pollution' (No. 37), accessed 3 September 2019 at http://www.gov.cn/zwgk/2013-09/12/content_2486773.htm.

State Council (2014), 'Circular on Disseminating the 2014–2015 Action Plan for Energy Conservation, Emissions Reduction and Low Carbon Development' (No. 23), accessed 9 April 2020 at http://www.gov.cn/zhengce/content/2014-05/26/content_8824.htm.

State Council (2016a), 'Circular of the State Council on Disseminating the National Strategic Emerging Industry Development Plan in the 13th Five Year Plan' (No. 67), accessed 15 July 2020 at http://www.gov.cn/zhengce/content/2016-12/19/content_5150090.htm.

State Council (2016b), 'Comprehensive Work Plan of Energy Conservation and Emissions Reductions in the 13th FYP Period' (No. 74), accessed 9 April 2018 at https://www.ndrc.gov.cn/gzdt/201701/t20170105_834501.html.

State Council (2016c), 'The 13th Five Year Plan for Economic and Social Development of the People's Republic of China (Outline)', accessed 9 April 2020 at http://www.xinhuanet.com/politics/2016lh/2016-03/17/c_1118366322.htm.

State Council (2016d), 'The Work Program for Controlling Greenhouse Gas Emissions in the 13th FYP Period', accessed 04 April 2018 at http://www.gov.cn/zhengce/content/2016-11/04/content_5128619.htm.

State Council (2020), 'China's Energy Development in the New Era (White Paper)', Beijing, accessed 25 December 2020 at http://www.gov.cn/zhengce/2020-12/21/content_5571916.htm.

State Council (2021), 'The 14th Five Year Plan for National Economic and Social Development of the People's Republic of China and the Outline of Long-Term Goals for 2035', Beijing, accessed 24 March 2021 at http://www.gov.cn/xinwen/2021-03/13/content_5592681.htm.

State Council General Affairs Office (SCGAO) (2006), 'Circular on Forwarding the Opinions of the National Development and Reform Commission on Improving the Policy of Differential Tariffs' (No.77), accessed 15 January 2020 at http://www.gov.cn/zwgk/2006-09/22/content_396258.htm.

State Council Press Office (SCPO) (2007), 'Press Conference on the System of Statistics and Monitoring for Energy Conservation and Emissions Reduction', accessed 11 April 2020 at http://www.china.com.cn/zhibo/2007-11/29/content_9304334.htm.

State Council Press Office (SCPO) (2018), 'Introducing China's Policy and Action in Addressing Climate Change Annual Report 2018', accessed 5 April 2020 at http://www.gov.cn/xinwen/2018-11/26/content_5343360.htm.

State Environmental Protection Administration (SEPA) (2008), 'Circular on Standardizing the Provision of Enterprise Environmental Violation Information to the Credit Reporting System of the People's Bank of China' (No. 33), accessed 26 April 2020 at http://www.csrcare.com/Law/Show?id=34218.

State Environmental Protection Administration (SEPA), The People's Bank of China (PBC) and China Banking Regulatory Commission (CBRC) (2007), 'Opinions on Implementing Environmental Protection Policies and Rules to Prevent Credit Risks' (No. 108), accessed 24 April 2020 at https://wenku.baidu.com/view/95fec19bdaef5ef7ba0d3c10.htm.

Stensdal, I. (2012), 'China's Climate-Change Policy 1988–2011: From Zero to Hero?', accessed 27 February 2020 at https://www.fni.no/getfile.php/131942-1469869900/Filer/Publikasjoner/FNI-R0912.pdf.

Stensdal, I. (2014), 'Chinese Climate-Change Policy, 1988–2013: Moving On Up', *Asian Perspective*, 38 (1), 111–35.

Stern, N. (2007), *The Economics of Climate Change: The Stern Review*, Cambridge: Cambridge University Press.

Su, W. (2010), 'China Copenhagen Accord Pledge, Correspondence to the UNFCCC Secretariat', accessed 14 August 2018 at https://unfccc.int/files/meetings/cop_15/copenhagen_accord/application/pdf/chinacphaccord_app2.pdf.

Su, W. (2015), 'China's Intended Nationally Determined Contribution', accessed 2 April 2020 at https://www4.unfccc.int/sites/submissions/indc/Submission Pages/submissions.aspx.

Su, X. (2011), 'Revolution and Reform: The Role of Ideology and Hegemony in Chinese Politics', *Journal of Contemporary China*, 20 (69), 307–26.

Syntao Green Finance (SGF) (2017), 'Semi Annual Report on the Incentive Mechanism of Green Bonds of Local Governments in China (the First Half of 2017)', accessed 26 April 2020 at http://syntaogf.com/Uploads/files/绿债2017半年报.pdf.

Syntao Green Finance (SGF) and Climate Bonds Initiative (CBI) (2017), 'Study of China's Local Government Policy Instruments for Green Bonds', accessed 26 April 2020 at https://www.climatebonds.net/files/reports/chinalocalgovt_02_13.04_final _a4.pdf.

Tanpaifang (2013), 'List of the Second Batch of Comprehensive Demonstration Cities of Energy Conservation and Emissions Reduction Fiscal Policy', accessed 27 November 2019 at http://www.tanpaifang.com/tanguwen/2013/1021/25129.html.

Tanpaifang (2019), 'In the First Quarter, a Total of 13.7277 Million Tons of Quota Were Sold in the National Pilot Carbon Market, with a Trading Volume of 263 Million Yuan', accessed 2 December 2019 at http://www.tanpaifang.com/tanjiaoyi/2019/0416/63590.html.

Tao, X., P. Yang, L. Ge, F. Cui, Q. Zhang, H. Wang, T. Peng and J. Shi (2013), 'A Study of the Technical Roadmap of Developing Low Carbon Economy in Shanghai', in *Research Report on Excellent Decision Consultation of Shanghai Association of Science and Technology*, Shanghai: Academia Press.

Task Force on Climate-Related Financial Disclosure (TCFD) (2017), *Final Report: Recommendations of the Task Force on Climate-Related Financial Disclosures* (June), Basel: Financial Stability Board. Downloadable from: https://www.fsb-tcfd.org/publications/.

Toke, D. (2017), *China's Role in Reducing Carbon Emissions*, London and New York: Routledge.

Torfing, J., B. G. Peters, J. Pierre and E. Sørensen (2012), *Interactive Governance: Advancing the Paradigm*, Oxford: Oxford University Press.

Tyfield, D. (2014), 'Putting the Power in "Socio-Technical Regimes": E-Mobility Transition in China as Political Process', *Mobilities*, 9 (4), 585–603.

UK-China Climate and Environmental Information Disclosure Pilot (UKCCEIDP) (2018), 'Report on the Progress of Pilot Program of UK-China Climate and Environmental Disclosure', accessed 28 April 2020 at https://www.unpri.org/climate-change/progress-report-on-uk-china-climate-and-environmental-disclosure-pilot/3875.article.

UK-China Climate and Environmental Information Disclosure Pilot (UKCCEIDP) (2020), *2019 Progress Report*. Downloadable from: https://www.unpri.org/climate-change/uk-china-pilot-on-climate-and-environmental-risk-disclosure-2nd-year-progress-report/5744.article.

UNEP (2019), *Emissions Gap Report 2019*, accessed 23 February 2020 at https://wedocs.unep.org/bitstream/handle/20.500.11822/30797/EGR2019.pdf.

UNEP Inquiry (2015), *The Financial System We Need: Aligning the Financial System with Sustainable Development*. The UNEP Inquiry Report, accessed 27 August 2020

at   https://unepinquiry.org/publication/inquiry-global-report-the-financial-system
-we-need/.
UNEP Inquiry/World Bank Group (2017), *Roadmap for a Sustainable Financial System*, accessed 13 April 2020 at http://unepinquiry.org/publication/roadmap-for-a
-sustainable-financial-system/.
UNFCCC (1997), 'Kyoto Protocol to the United Nations Framework Convention on Climate Change', accessed 17 May 2020 at https://unfccc.int/documents/2409.
UNFCCC (2015), 'Adoption of the Paris Agreement', accessed 14 August 2018 at https://unfccc.int/resource/docs/2015/cop21/eng/l09r01.pdf.
United Nations (UN) (1992), 'United Nations Framework Convention on Climate Change', accessed 1 April 2020 at https://unfccc.int/process-and-meetings/the
-convention/what-is-the-united-nations-framework-convention-on-climate-change.
Victor, D. G., F. W. Geels and S. Sharpe (2019), 'Accelerating the Low Carbon Transition: The Case for Stronger, More Targeted and Coordinated International Action', accessed 27 August 2020 at http://www.energy-transitions.org/content/
accelerating-low-carbon-transition.
Wan, X. (2016), 'Governmentalities in Everyday Practices: The Dynamic of Urban Neighbourhood Governance in China', *Urban Studies*, 53 (11), 2330–46.
Wan, Z., D. Sperling and Y. Wang (2015), 'China's Electric Car Frustrations', *Transportation Research Part D: Transport and Environment*, 34, 116–21.
Wang, A. (2014), 'Qingdao Energy Project Won the First ADB Loan of US $130 Million', accessed 9 March 2014 at http://news.qingdaonews.com/qingdao/2014-04/
11/content_10384610.htm.
Wang, A. L. (2013), 'The Search for Sustainable Legitimacy: Environmental Law and Bureaucracy in China', *Harvard Environmental Law Review*, 37, 365–440.
Wang, F., S. Yang, A. Reisner and N. Liu (2019), 'Does Green Credit Policy Work in China? The Correlation between Green Credit and Corporate Environmental Information Disclosure Quality', *Sustainability*, 11 (3), 733.
Wang, K., Q. Xue and Y. Y. Che (2016), 'A Study of Low Carbon Development Pathways and Carbon Emission Peak in Qingdao', in Green Low-Carbon Development Think-Tank Partnership (GLCDTTP) (ed.), *Policies and Practices in China's Low-Carbon Pilot Cities: Bring the Emissions Peak Forward*, Beijing: Science Press, pp. 92–111.
Wang, Y. (2015), 'The Rise of the "Shareholding State": Financialization of Economic Management in China', *Socio-Economic Review*, 13 (3), 603–25.
Wang, Y. (2018), 'Green Finance: The Progress Status and Research in China and the World'. Lecture at the Institute for Sustainable Resources, UCL, London, 31 January.
Wang, Y., Q. Song, J. He and Y. Qi (2015), 'Developing Low-Carbon Cities through Pilots', *Climate Policy*, 15 (supp. 1), S81–S103.
Wang, Y. and S. Zadek (2017), 'Establishing China's Green Financial System: Progress Report 2017 Summary', accessed 21 April 2020 at http://unepinquiry
.org/wp-content/uploads/2017/11/China_Green_Finance_Progress_Report_2017
_Summary.pdf.
Wang, L. and X. Zhang (2019), 'Making an Urban Ecotopia in China: Knowledge, Power, and Governmentality', in X. Zhang (ed.), *Remaking Sustainable Urbanism: Space, Scale and Governance in the New Urban Era*, London: Palgrave Macmillan, pp. 37–55.

Water 8848 (2014), 'Sino French Water Limited Signed Sewage Treatment Contract with Qinggang', accessed 9 March 2021 at http://www.water8848.com/news/201401/23/13413.html.

Weng, Y., D. Zhang, L. Lu and X. Zhang (2018), 'A General Equilibrium Analysis of Floor Prices for China's National Carbon Emissions Trading System', *Climate Policy*, 18 (supp. 1), S60–S70.

Weng, Y. Y., T. Y. Qi and X. L. Zhang (2017), 'Low-Carbon Energy Transition Outlook in China', in X. L. Zhang and Y. Qi (eds), *Annual Review of Low-Carbon Development in China 2017*, Beijing: Social Sciences Academic Press (China), pp. 73–88.

Westman, L. and V. Castan Broto (2018), 'Climate Governance through Partnerships: A Study of 150 Urban Initiatives in China', *Global Environmental Change*, 50, 212–21.

Wharton School (2019), *China's Electric Vehicle Market: A Storm of Competition Is Coming*, accessed 7 July 2020 at https://knowledge.wharton.upenn.edu/article/chinas-ev-market/.

Whiting, S. (2004), 'The Cadre Evaluation System at the Grass Roots: The Paradox of Party Rule', in B. Naughton and D. Young (eds), *Holding China Together: Diversity and National Integration in the Post-Deng Era*, Cambridge: Cambridge University Press, pp. 101–19.

Wieczorek, A. J. (2018), 'Sustainability Transitions in Developing Countries: Major Insights and their Implications for Research and Policy', *Environmental Science and Policy*, 84, 204–16.

World Bank (2020), *State and Trends of Carbon Pricing 2020*, Washington, DC: World Bank.

World Commission on Environment and Development (WCED) (1987), *Our Common Future*, Oxford: Oxford University Press.

World Meteorological Organization (WMO), UN Environment (UNEP), Intergovernmental Panel on Climate Change (IPCC), Global Carbon Project, Future Earth, Earth League and the Global Framework for Climate Services (GFCS) (2019), *United in Science: High-Level Synthesis Report of Latest Climate Science Information Convened by the Science Advisory Group of the UN Climate Action Summit 2019*, accessed 17 May 2020 at https://public.wmo.int/en/resources/united_in_science.

World Resources Institute (WRI) (2011a), 'Phase I of Qingdao City Jiazhou Bay Water Quality Protection Project', Inception Report of TA No. 7219-PRC to the Asian Development Bank, Washington, DC: World Resources Institute.

World Resources Institute (WRI) (2011b), 'Phase II of Qingdao City Jiazhou Bay Water Quality Protection Project', Interim Report of TA No. 7219-PRC to the Asian Development Bank, Washington, DC: World Resources Institute.

World Resources Institute (WRI) (2011c), 'Press Release: Multinational Corporations Should Play an Active Role in the Construction of Low Carbon Cities in China – Qingdao Seminar on the Role of Multinational Companies in Low Carbon Development Successfully Held', accessed 10 March 2020 at http://www.wri.org.cn/en/node/40806.

World Resources Institute (WRI) (2012), Phase II of Qingdao City Jiazhou Bay Water Quality Protection Project', Draft Final Report of TA No. 7219-PRC to the Asian Development Bank, Washington, DC: World Resources Institute.

Wu, L. P., H.-J. Jian and Y. Zhou (2016), 'A Study of Shanghai Green Transition and Development (2015–2030)', in Green Low-Carbon Development Think-Tank

Partnership (GLCDTTP) (ed.), *Policies and Practices in China's Low-Carbon Pilot Cities: Bringing the Emissions Peak Forward*, Beijing: Science Press, pp. 135–64.

WWF Shanghai Low Carbon Development Roadmap Research Team (WWF-SLCDRRT) (ed.) (2011), *2050 Shanghai Low Carbon Development Roadmap Report*, Beijing: China Science Publishing & Media Ltd.

Xi, J. (2007), *Zhejiang, China: A New Vision for Development*, Hangzhou: Zhejiang United Press and Zhejiang Renmin Press.

Xi, J. (2014), 'Speech by H.E. Xi Jinping President of the People's Republic of China at the Körber Foundation', accessed 30 March 2020 at https://www.fmprc.gov.cn/mfa_eng/wjdt_665385/zyjh_665391/t1148640.shtml.

Xi, J. (2017a), 'Securing a Decisive Victory in Building a Moderately Prosperous Society in All Respects and Winning the Great Victory of Socialism with Chinese Characteristics in the New Era', Report to the 19th Congress of the CPC on 18 October 2017, accessed 29 March 2020 at http://www.xinhuanet.com//politics/19cpcnc/2017-10/27/c_1121867529.htm.

Xi, J. (2017b), *The Governance of China*, Beijing: Foreign Languages Press.

Xi, J. (2020), 'Full Text: Remarks by Chinese President Xi Jinping at Climate Ambition Summit', accessed 8 April 2021 at http://www.xinhuanet.com/english/2020-12/12/c_139584803.htm.

Xiao, J. and X. Zhou (2013), 'An Empirical Assessment of the Impact of the Vehicle Quota System on Environment: Evidence from China', accessed 28 August 2019 at http://www.econ.cuhk.edu.hk/econ/zh-tw/news-event/news-media?seminar_searchbox=8.

Xiao, J., X. Zhou and W.-M. Hu (2017), 'Welfare Analysis of the Vehicle Quota System in China', *International Economic Review*, 58 (2), 617–50.

Xinhua Finance (2016), '16 Bank of Qingdao Green Finance 01', accessed 8 March 2021 at http://greenfinance.xinhua08.com/a/20160525/1641453.shtml?f=arelated.

Xinhua Finance (2019), 'The New International Green Bond Standard Has Been Issued and the "International Compliance Rate" of China's Green Bonds Needs to Be Improved', accessed 8 March 2021 at http://greenfinance.xinhua08.com/a/20191216/1902824.shtml?ulu-rcmd.

Xinhuanet (2013), 'Xi Jinping: Ideological Work Is an Extremely Important Task for the Party', accessed 31 July 2019 at http://www.xinhuanet.com/politics/2013-08/20/c_117021464_3.htm.

Xu, C. (2011), 'The Fundamental Institutions of China's Reform and Development', *Journal of Economic Literature*, 49 (4), 1076–1151.

Xu, C. (2017), 'The Pitfalls of a Centralized Bureaucracy: Acceptance Speech for 2016 China's Economics Prize', accessed 24 August 2020 at http://www.fmsh.fr/en/college-etudesmondiales/30088.

Xu, G. Q., X. L. Hu, S. Y. Weng and Y. P. Hu (2016), 'Coal Consumption Scenarios for the 293 Prefecture-level Municipalities in China (2010–2050)', in Green Low-Carbon Development Think-Tank Partnership (GLCDTTP) (ed.), *Policies and Practices in China's Low-Carbon Pilot Cities: Bringing the Emissions Peak Forward*, Beijing: Science Press, pp. 236–56.

Xu, J. (2016), 'Environmental Discourses in China's Urban Planning System: A Scaled Discourse-Analytical Perspective', *Urban Studies*, 53 (5), 978–99.

Xu, J. and C. Chung (2014), '"Environment" as an Evolving Concept in China's Urban Planning System', *International Development Planning Review*, 36 (4), 391–412.

Xu, N. (2019), 'What Gave Rise to China's Land Finance?', *Land Use Policy*, 87, 104015.

Xue, L. and H. Zhang (2014), 'Low-Carbon and Sustainable Transport for Qingdao: A Strategic Study (Full Report in Chinese)', in World Research Institute (ed.), *Sustainable and Liveable Cities Initiative for Qingdao*, Beijing: World Research Institute.

Xue, Y., J. You, X. Liang and H.-C. Liu (2016), 'Adopting Strategic Niche Management to Evaluate EV Demonstration Projects in China', *Sustainability*, 8 (2), 142.

Yang, J., Y. Liu, P. Qin and A. A. Liu (2014), 'A Review of Beijing's Vehicle Registration Lottery: Short-Term Effects on Vehicle Growth and Fuel Consumption', *Energy Policy*, 75, 157–66.

Yu, C.-J. and R. Li (2017), 'The Current State of Green Bonds Information Disclosure and Suggestions', accessed 22 April 2020 at http://www.360doc.com/content/18/0320/16/39103730_738752209.shtml.

Yu, X.-G., Y. Lin and Y.-X. Chen (2013), *Green Credit Footprint of China's Banking Industry* (*Zhongguo Yinhangye Xindai Zuji*), Beijing: China Environment Press.

Yu, Z.-B. and Z.-M. Meng (2017), 'Some Thoughts on Energy Use Allowances Trading', accessed 1 April 2020 at http://www.tanjiaoyi.com/article-21418-1.html.

Yuan, J., Y. Xu and Z. Hu (2012), 'Delivering Power System Transition in China', *Energy Policy*, 50, 751–72.

Yuan, X. and J. Zuo (2011), 'Transition to Low Carbon Energy Policies in China: From the Five-Year Plan Perspective', *Energy Policy*, 39 (6), 3855–9.

Zadek, S. and C. Zhang (2014), 'Greening China's Financial System: An Initial Exploration', accessed 25 August 2020 at https://www.iisd.org/publications/greening-chinas-financial-system-initial-exploration.

Zhang, B., Y. Yang and J. Bi (2011), 'Tracking the Implementation of Green Credit Policy in China: Top-Down Perspective and Bottom-Up Reform', *Journal of Environmental Management*, 92 (4), 1321–7.

Zhang, C., S. Zadek, N. Chen and M. Halle (2015), 'Section 1: Synthesis', in International Institute for Sustainable Development (IISD) and Development Research Center (DRC) of the State Council (eds.), *Greening China's Financial System*, Winnipeg: IISD, pp. 1–26.

Zhang, J., Z. Wang and X. Du (2017), 'Lessons Learned from China's Regional Carbon Market Pilots', *Economics of Energy & Environmental Policy*, 6 (2), 19–38.

Zhang, L., A. P. J. Mol and D. A. Sonnenfeld (2007), 'The Interpretation of Ecological Modernisation in China', *Environmental Politics*, 16 (4), 659–68.

Zhang, L. and L. Zhao (2010), 'Historical Interaction between Environmental Conceptions and Man-Land Relationship since the Establishment of New China', *Researches in Chinese Economic History*, 1, 3–11.

Zhang, L.-Y. (2003), 'Economic Development in Shanghai and the Role of the State', *Urban Studies*, 40 (8), 1549–72.

Zhang, L.-Y. (2004), 'The Roles of Corporatization and Stock Market Listing in Reforming China's State Industry', *World Development*, 32 (12), 2031–47.

Zhang, L.-Y. (2014), 'Dynamics and Constraints of State-led Global City Formation in Emerging Economies', *Urban Studies*, 51 (6), 1162–78.

Zhang, L.-Y. (2015), 'Rethinking China's Low-Carbon Strategy', accessed 12 April 2020 at http://www.paulsoninstitute.org/wp-content/uploads/2017/01/PPM_Chinas-Low-Carbon-Strategy_Zhang_English_R.pdf.

Zhang, L.-Y. (2019), 'Green Bonds in China and the Sino-British Collaboration: More a Partnership of Learning than Commerce', *British Journal of Politics and International Relations*, 21 (1), 207–25.

Zhang, Y. (2009), 'Liu Zongchao: Singing in the Time and Space of Ecological Civilization', accessed 12 April 2020 at http://finance.sina.com.cn/leadership/mroll/20091123/16377003618.shtml.

Zhang, Z. (2018), 'Energy Price Reform in China', in R. Garnaut, L. Song and C. Fang (eds.), *China's 40 Years of Reform and Development 1978–2018*, Canberra: ANU Press, pp. 507–24.

Zhao, X. and W. Li (2018), 'Targets Management in China's Forty-Year Energy Conservation Policy', in Y. Qi and X. Zhang (eds.), *Annual Review of Low-Carbon Development in China*, Beijing: Social Sciences Academic Press (China), pp. 24–50.

Zhao, X., M. Tie, J. Zhang and Z. Yu (2018), 'Historical Events of China's Energy Conservation in Forty Years', in Y. Qi and X. Zhang (eds.), *Annual Review of Low-Carbon Development in China*, Beijing: Social Sciences Academic Press (China), pp. 1–23.

Zhao, Z. (1987), 'Zhao Ziyang's Report at the 13th Congress of the CPC, 25 Oct. 1987', accessed 26 November 2018 at http://cpc.people.com.cn/GB/64162/64168/64566/65447/4526368.html.

Zhejiang Center for Climate Change and Low-Carbon Development Cooperation (ZCCCLCDC) (2015), 'Liang Shen Ding Tan: Zhejiang as the First Province to Routinise the Compilation of Greenhouse Gases Inventory on Three Levels Covering Province-Municipality-County', Beijing: World Resources Institute.

Zheng, L. (2018), 'The Macro Prudential Assessment Framework of China: Background, Evaluation and Current and Future Policy', CIGI Papers No. 164, Waterloo, accessed 5 March 2021 at https://www.cigionline.org/publications/macro-prudential-assessment-framework-china-background-evaluation-and-current-and.

Zheng, S., M. E. Kahn, W. Luo and D. Sun (2013), *Incentivizing China's Urban Mayors to Mitigate Pollution Externalities: The Role of the Central Government and Public Environmentalism*, Cambridge, MA: National Bureau of Economic Research. Downloadable from: https://www.nber.org/papers/w18872.pdf.

Zheng, Z. (2015), 'Demand for Green Finance', in International Institute for Sustainable Development (IISD) and Development Research Center (DRC) of the State Council (eds.), *Greening China's Financial System*, Winnipeg: IISD, pp. 44–61.

Zhihu (2018), 'The Development History of Shanghai's Vehicle License Auction', accessed 28 August 2020 at https://zhuanlan.zhihu.com/p/34293518.

Zhou, E. (1954), 'Report on the Work of the Government of the State Council in 1954', accessed 26 November 2018 at http://www.gov.cn/test/2006-02/23/content_208673.htm.

Zhou, E. (1975), 'Report on the Work of the Government of State Council 1975', accessed 26 November 2018 at http://www.gov.cn/test/2006-02/23/content_208796.htm.

Zhou, J.-L. (2010), 'Qingdao: Developing a Low-Carbon Economy by World Standards', accessed 13 May 2020 at http://finance.sina.com.cn/chanjing/b/20101117/10128965967.shtml.

Zhu, B. (2017), 'Fang Xinghai: China Securities Regulatory Commission Is Studying to Extend Mandatory Environmental Disclosure to All Listed Companies', accessed 8 March 2021 at http://www.greenfinance.org.cn/displaynews.php?id=772.

Zhu, Z. X. (2018), 'Reform the Planning System and Give Better Play to the Role of Planning', accessed 21 July 2020 at https://www.sohu.com/a/271721997_99917590.

Zhuo, X. and L. Zhang (2015), 'A Framework for Green Finance: Making Clear Waters and Green Mountains China's Gold and Silver', in International Institute for Sustainable Development (IISD) and Development Research Center (DRC) of the State Council (eds.), *Greening China's Financial System*, Winnipeg: IISD, pp. 28–43.

Zou, H., H. Du, J. Ren, B. K. Sovacool, Y. Zhang and G. Mao (2017), 'Market Dynamics, Innovation, and Transition in China's Solar Photovoltaic (PV) Industry: A Critical Review', *Renewable and Sustainable Energy Reviews*, 69, 197–206.

Zuo, C. (Vera) (2015), 'Promoting City Leaders: The Structure of Political Incentives in China', *The China Quarterly*, 224, 955–84.

Index